Human Factors and Decision Making:
Their Influence on Safety and Reliability

BPS
ES
ESRA
HFRG
I Chem E
IEE
UKSS

SARSS '88

Proceedings of the Safety and Reliability Society Symposium, 1988, held at Altrincham, Manchester, UK, 19–20 October 1988

Organised by
The Safety and Reliability Society, Clayton House, 59 Piccadilly, Manchester M1 2AQ, UK

Co-sponsors
The British Psychological Society
The Ergonomics Society
The European Safety and Reliability Association
The Human Factors in Reliability Group
The Institution of Chemical Engineers
The Institution of Electrical Engineers
The UK Systems Society

Conference and Symposium Executive
Dr G. B. Guy (Chairman)
Mr E. Johnson (Secretary)
Ms B. A. Sayers (Editor)

Mr R. F. Cox
Mr G. Turner
Dr M. H. Walter

Mr F. Oakes (IEE representative)
Mr I. A. Watson (HFRG representative)

Paper Selection and Review Panel
Panel drawn from members of The Human Factors in Reliability Group (HFRG)

Human Factors and Decision Making:
Their Influence on Safety and Reliability

Edited by

BEVERLEY A. SAYERS

RM Consultants Ltd, Warrington, UK
and
Safety and Reliability Society, Manchester, UK

ELSEVIER APPLIED SCIENCE
LONDON and NEW YORK

ELSEVIER SCIENCE PUBLISHERS LTD
Crown House, Linton Road, Barking, Essex IG11 8JU, England

Sole Distributor in the USA and Canada
ELSEVIER SCIENCE PUBLISHING CO., INC.
52 Vanderbilt Avenue, New York, NY 10017, USA

WITH 38 TABLES AND 53 ILLUSTRATIONS

© 1988 ELSEVIER SCIENCE PUBLISHERS LTD
© 1988 CROWN COPYRIGHT—pp. 51–70
© 1988 UNITED KINGDOM ATOMIC ENERGY AUTHORITY—pp. 71–86, 110–125, 202–216
© 1988 J. C. WILLIAMS—pp. 223–240
Softcover reprint of the hardcover 1st edition 1988

British Library Cataloguing in Publication Data

Safety and Reliability Society, *Symposium*
(*1988: Altrincham, England*)
Human factors and decision making, their
influence on safety and reliability.
1. Industrial safety. Management aspects
I. Title II. Sayers, Beverley A.
658.3'82

Library of Congress Cataloging in Publication Data

Human factors and decision making.

Bibliography: p.
1. Human engineering. 2. Industrial safety.
3. Reliability (Engineering) I. Sayers, Beverley A.
T59.7.H84 1988   620.8'2   88-24437

ISBN-13: 978-94-010-7113-0   e-ISBN-13: 978-94-009-1375-2

DOI: 10.1007/978-94-009-1375-2

No responsibility is assumed by the Publisher for any injury and/or damage to persons or property as a matter of products liability, negligence or otherwise, or from any use or operation of any methods, products, instructions or ideas contained in the material herein.

Special regulations for readers in the USA

This publication has been registered with the Copyright Clearance Center Inc. (CCC), Salem, Massachusetts. Information can be obtained from the CCC about conditions under which photocopies of parts of this publication may be made in the USA. All other copyright questions, including photocopying outside the USA, should be referred to the publisher.

All rights reserved. No part of this publication may be reproduced, stored in a retrieval system, or transmitted in any form or by any means, electronic, mechanical, photocopying, recording, or otherwise, without the prior written permission of the publisher.

P R E F A C E

The recent major disasters at Bhopal, Chernobyl and Zeebrugge and the human failures which are seen to have led to such loss of life have had a great impact on the public. This has resulted in a general awareness of the dependence of safety on the technical ability and reliability of the personnel involved in the design, operation or maintenance of plants and equipment. It is evident that human reliability is too often (some would argue always) the weak link in the chain of events which leads to loss of safety and catastrophic failure of even the most advanced technological systems.

This book represents the proceedings of the 1988 Safety & Reliability Symposium held in Altrincham, UK, from the 19th to 20th October 1988. It is thus part of the series of proceedings of society events and the reader should note that aspects of human performance were discussed and later reported after the Society's 1985 and 1986 symposia. The subject is so important that it is the Society's intention to continually review factors affecting human performance and to report on findings at future symposia.

The structure of the book represents the structure of the Symposium itself. The session titles and the papers as selected represent current thinking within many industries on the management of the risks inevitable in any man/machine system. In particular, attention is drawn to various techniques and methods of evaluating human reliability including the use of computer codes and data bases. Management factors, maintenance and the role of technical documentation are highlighted in several papers. Also addressed is the very real problem of emergency planning, which must be undertaken within any high risk industry, but particularly where an accident may affect the civilian population.

The response to the Call for Papers produced many more good papers than could be included in the programme. I must thank all the authors who submitted their work, the review panel (drawn from members of The Human Factors in Reliability Group (HFRG)), the presenters of the papers, the Symposium organising committee and the Co-sponsors for their support. Only through the hard work and dedication of these people was it possible to achieve such a successful Symposium and to compile this book.

B A Sayers

R M Consultants
Genesis Centre
WARRINGTON

Contents

Preface v

List of Contributors ix

HUMAN FACTORS IN SAFETY

Human Factors in the Process Industries 1
    R Grollier Baron

Human Factors Considerations in the Safety Case 10
for the Channel Tunnel Project
    A J Smith

Safety and Human Factors in Manned Space Flight Systems 23
    I Jenkins

The Role of Technical Documentation for Maintenance Staff in the Safe
Operation and Maintenance of Machine and Processes 39
    K Clements-Jewery

Allocation of Function Between Man and Programmable Electronic Systems
in Safety-Related Applications 51
    J Brazendale

HUMAN RELIABILITY ASSESSMENT

Human Reliability Assessors Guide : An Overview 71
    P Humphreys

A Comparative Evaluation of Five Human Reliability Assessment Techniques 87
    B Kirwan

Human Factors Reliability Benchmark Exercise, Report of the SRD 110
Participation
    T Waters

The Application of the Combined THERP/HCR Model in Human Reliability 126
Assessment
    R B Whittingham

A Practical Application of Quantified Risk Analysis 139
    G Purdy

PROF a Computer Code for Prediction of Operator Failure Rate 158
    K H Drager, H S Soma, O Falmyr

New Directions in Qualitative Modelling. 170
    D Lucas, D E Embrey

HUMAN FACTORS IN MANAGEMENT AND DESIGN

Management Factors and System Safety 172
    S Whalley, D Lihou

Addressing Human Factors Issues in the Safe Design and Operation of
Computer Controlled Process Systems 189
    L J Bellamy, T A W Geyer

Management in High Risk Industries 202
    I A Watson, F Oakes

Processor-Based Displays: The Flexible Control Panel 217
    T F Mayfield

A Human Factors Data-Base to Influence Safety and Reliability 223
    J C Williams

DECISION SUPPORT SYSTEMS

Modelling the Evacuation of the Public in the Event of Toxic Releases: A
Decision Support Tool and Aid for Emergency Planning 241
    P I Harrison, L J Bellamy

Designing Decision Support Systems for Human Error Reduction: The Need
to Address Information Distortion 256
    W L Cats-Baril

Supporting Expert Judgement of Human Performance and Reliability 270
    D E Embrey

List of Contributors

L J Bellamy
Technica Ltd, Lynton House, 7/12 Tavistock Square, London, WClH 9LT, UK.

J Brazendale
Control Systems Group, Technology Division, Health and Safety Executive, Magdalen House, Stanley Precinct, Bootle, Merseyside, L20 3QZ, UK.

W L Cats-Baril
School of Business Administration, University of Vermont, Burlington, Vermont 05405, U.S.A.

K Clements-Jewery
Assistant Head of School of Engineering, Glasgow College, Cowcaddens Road, Glasgow, G4 OBA, UK.

K H Drager
Managing Director, A/S Quasar Consultants, Munkedamsveien 53 B, 0250 Oslo 2, Norway.

D E Embrey
Human Reliability Associates, 1 School House, Higher Lane, Dalton, Wigan, Lancashire, WN8 7RP, UK.

O Falmyr
Statoil, Postboks 300, 4001 Stavanger, Norway.

T A W Geyer
Technica Ltd, Lynton House, 7/12 Tavistock Square, London, WClH 9LT, UK.

R Grollier Baron
Institut Francaise du Petrole, Centre d'Etudes et de Developpement Industriels, B.P. 3 - 69390 Vernaison, France.

P I Harrison
Technica Ltd, Lynton House, 7/12 Tavistock Square, London, WClH 9LT, UK.

P Humphreys
Safety and Reliability Directorate, United Kingdom Atomic Energy Authority, Wigshaw Lane, Culcheth, Warrington, WA3 4NE, UK.

I Jenkins
Messerschmitt-Bolkow-Blohm GmbH, Space Communications and Propulsion Division, Munich, Federal Republic of Germany.

B Kirwan
British Nuclear Fuels plc, Risley, Warrington, WA3 6AS, UK.

D Lihou
Lihou Loss Prevention Services Ltd, Grays Court, 1 Nursery Road, Edgbaston, Birmingham, B15 3JX, UK.

D Lucas
Human Reliability Associates, 1 School House, Higher Lane, Dalton, Wigan, Lancashire, WN8 7RP, UK.

T F Mayfield
Electrical and Electronics Design Department, Rolls-Royce and Associates Ltd, P O Box 31, Derby, DE2 8BJ, UK.

F Oakes
Engineering and Management Consultant, 3 Rollswood Road, Welwyn, Herts, AL6 9TX

G Purdy
Process Safety Group, Technology Division, Health and Safety Executive, Magdalen House, Stanley Precinct, Bootle, L20 3QZ, Merseyside, UK.

A J Smith
WS Atkins Engineering Sciences, Stanford House, Birchwood, Warrington, Cheshire, WA3 7BH, UK.

H S Soma
A/S Quasar Consultants, Munkedamsveien 53 B, 0250 Oslo 2, Norway.

T Waters
Reliability Technology Research Unit. National Centre of Systems Reliability, Safety and Reliability Directorate, UKAEA, Wigshaw Lane, Culcheth, Warrington, WA3 4NE, UK.

I A Watson
Head of Systems Reliability Service, United Kingdom Atomic Energy Authority, Wigshaw Lane, Culcheth, Warrington, WA3 4NE, UK.

S Whalley
Lihou Loss Prevention Services Ltd, Grays Court, 1 Nursery Road, Edgbaston, Birmingham, B15 3JX, UK.

R B Whittingham
Electrowatt Engineering Services (UK) Ltd, Grandford House, 16 Carfax, Horsham, West Sussex, RH12 1UP, UK.

J C Williams
Sizewell 'B' Project Management Team, Central Electricity Generating Board, Booths Hall, Knutsford, Cheshire, WA16 8QG, UK.

HUMAN FACTORS IN THE PROCESS INDUSTRIES

Roger GROLLIER BARON
INSTITUT FRANCAIS DU PETROLE
Centre d'Etudes et de Développement Industriels
B.P. 3 - 69390 VERNAISON - France

ABSTRACT

The aim of this paper is to review techniques for reducing the share of the human factor in causes of accidents. Although the separation is not always clearcut, such techniques have been classified in three categories:

- the ones in which the operator is adapted to his workstation with regard to safety and quality;

- the ones in which the system and organization are adapted to operators, with these latter supposed to show proof of proper behavior;

- the ones ensuring that overall man/system performances are on a good level with regard to safety and quality.

The implementation of all or some of the techniques described there results in a great improvement in performances with respect to safety and quality, hence in the economies of the installation.

INTRODUCTION

It is well known that human factors play a primordial role in accidents. It is assumed, on the average, that they make up 80% of the causes leading to accidents in the processing industries. This is why numerous studies have been carried out to better understand the origin of such human failures and to correct them.

Without going into details about these studies, we will limit ourselves to giving the principal practical conclusions leading to great improvements in this field. They can be classified in two groups: (1) those concerning the adapting of a worker to his work-station (psychosociological approach); (2) those concerning the adapting of the workstation to the worker (scientific approach).

1° - Adapting a Worker to His Workstation

It is known that safety begins by the strict application of well-conceived procedures. This involves: (1) the state of mind of the operator, who must be motivated; (2) the aptitude of the operator.

1.1 - Motivation

Motivation begins by making the operator aware of the importance of the problem. This is the role of any hierarchy, beginning with the manager who, by his example, must show the importance that the management attaches to safety.

This must be completed by encouragement, which may take on several forms:

- personal involvement in the actions of the operator, e.g. his signature on check lists;

- delegating responsibility and participating in decisions;

- culture of the establishment in which personnel are assimilated into the establishment;

- dissuasion while taking into consideration safety concerns in staff appreciation ;

- social climate, work ambience and social pressure;
  . orderliness,
  . cleanliness,
  . and also the teamwork context.

All these factors may encourage operators to have a stricter behavior.

1.2 - Aptitude

During recruitment, the spirit of discipline should be preferred rather than boldness for personnel called upon to operate dangerous installations. The physiological condition of such persons must also make them more proficient in doing their job, in particular in degraded situations. Among the important parameters are: (1) sensitivity to stress, (2) alcoholism, (3) use of tranquilizers, and (4) drugs.

Aptitude requires training to know the installation well and to face up to dangerous situations. This latter factor is taking on more and more importance because installations are increasingly automated and reliable. Potential dangerous situations have a slight probability of occurring, and the operator must be capable of controlling them.

This latter aptitude must be checked about every six months. The development of simulators should facilitate this checking.

Lacking a simulator, we can imagine dangerous transitory situations from fault trees compiled during safety analyses, and we can check to see that operators of the proper reactions when faced with the situations with which they are confronted.

Operators running dangerous installations should have a skill with a limited duration (e.g. one year), but which is renewable after checking to see that the aptitude has been maintained.

2. Adapting the Workstation to the Worker

The mental and physical work load that can be required of a worker facing a degraded situation is limited. This limitation is poorly understood and varies from one individual to another and from one instant to another for the same person. Hence the acceptable work load must be optimized by being reduced.

2.1 Optimizing the Acceptable Work Load

This implies that the operator is able to concentrate on his job, i.e.:

- by being in a comfortable material situation and protected from outside aggressions (noise, vibrations, temperature, etc.),

- by not being stressed by outside conditions, i.e. finding himself in a secure place whatever the outside conditions may be (explosion, toxic cloud, etc.), i.e. in a control room suitably resistant to outside aggressions.

2.2. Reducing the Work Load

This implies that the operator easily has all useful indications for controlling the situation and only these indications.

The tendency is to separate safety parameters from management parameters in control rooms. The operator must be able, at all times, to follow the evolution of all essential safety parameters in an analog and not digital form. He must not have to call up the evolution of any given safety parameter on the screen. Information must be easy to interpret and be presented logically.

Redundant information must be eliminated from safety parameters. This one has to be found among the management parameters. In particular, a distinction must be made between important alarms and the ensuing secondary alarms. The former must be forceful to make up for losses of vigilance, whether audible or visual (e.g. flashing alarm lights), while the latter must be much more discrete (winking only).

Controls must be arranged unambiguously.

2.3. Procedures

Procedures are an important aspect. A distinction must be made between voluminous general documents explaining how the installation operates and actual procedures giving the detailed actions to be undertaken, either constantly or when faced with exceptional situations.

2.3.1 - Procedural Characteristics in Common

- Procedures must be clear, precise and understandable for all operators involved. It is thus recommended that they be written in close participation with the persons who must implement them, so that the language used does not give rise to a different interpretation.

It is also an opportunity to see that models representing the system, according to which operators work out their strategy in the case of disfunctioning, are correct.

- The parts to be manipulated must be designated correctly to prevent any confusion. It is best to accompany procedures with diagrams in which these parts are indicated. The text of the procedure must indicate the function and reference of these parts. The reference must be indicated on the part itself.

Concerning this, it is important to avoid confusion with parts that are operated only exceptionally, for maintenance, for example. It is recommended that these latter be sealed with lead in their normal position. In case of urgent necessity, it is still possible to operate them without losing any time hunting for a key that may be lost. The affixing of padlocks or key-operated systems must be used for maintenance work and for consignment of lines or equipment when the operational personnel might be victim of an untimely start-up.

- The procedures must specify the jobs and responsibilities of each as well as the coordination required.

- The procedures must draw attention to what must not be done.

- The procedures must be easy to use when the time comes:

  . easy to read: no small print, few sentences, use of colors and diagrams;
  . rugged support (e.g. plasticized paper);
  . ease of access (classification, availability, kept in a sealed plastic bag for emergency procedures, etc.).

- The procedures must explain the means of protection to be used when combatting emergencies (protective clothing, safety parts, etc.).

2.3.2 - Routine Procedures

These procedures are mainly for batch manufacturing operations, start-ups and shutdowns, scheduled or not, of installations operating continuously.

For these procedures, it is recommended that check-lists be used, on which the operators check off the operations progressively as they are done. This avoids forgetting some jobs. If the operator is inconveniently unavailable, his replacement knows the exact degree of advance of the process.

Procedures on a check-list are made mandatory by some quality-control agencies when product quality is an essential criterion (e.g. pharmaceutical base stocks, special alloys).

Naturally, the check-list can be displayed on a screen where the operations performed can be ticked off.

It should be noted that many accidents stem from not strictly respecting procedures. Some operators believe they can simplify work or save time by not doing some jobs that are actually redundant. For example, they may not check to see whether a parameter has reached a given value. If a breakdown occurs, the redundancy is then no longer available to handle the situation. This is an additional reason for writing out procedures with the operators, while seeing that they understand the importance of apparently superfluous jobs.

In the check-lists, it is best to include the steady states, i.e. situations that can be maintained for a fairly long time with full safety e.g. the total reflux for a distillation column. These are states towards which it may be possible to return in case of difficulty in subsequent operations.

Likewise, it should be mentioned what is to be done if an operation fails.

2.3.3 - Procedures Relating to Abnormal Situations

The problem is to help operators to face up to an unusual situation that may lead to an accident or to losses. Such situations are revealed during a safety analysis. The corresponding procedures aim, on the basis of data observed by the operator, to guide him in his diagnosis and to help him work out his strategy and define the objectives to be reached. Such procedures must, first of all, indicate the additional information to be obtained and the checks to be made to fully understand the causes of disfunctioning and to deduce what has to be done.

The operator can be helped by including various data such as curves relating to some parameters. Such data must also serve as a warning with regard to dangerous situations or operations.

Mention must also be made of the personnel means required and the documents to which reference must be made.

Depending on the situation and the qualification of operators, a procedure can be worked out to bring the workstation back to the authorized domain, hence to the steady state, or simply to trigger an emergency shutdown.

To the extent where progress in automation has made possible the monitoring of a greater number of installations by a single person, together with the greater operational reliability of such installations, hence meaning that the operator will have little opportunity to intervene and the phenomena occurring have fast kinetics, it may be safer to plan on triggering an emergency shutdown.

3° - Maintenance and Modifications

The rigor required in operating installations must be reflected in the maintenance and modification procedures.

In particular, no work, even of a modest and harmless appearance, must be required without being checked by the management, or even by the safety services.

For operations occurring with some frequency, check-lists can be compiled.

Concerning modifications, care must be taken to see that the change planned does not create any additional risk. This requires the existence of instructions or of a procedure as to how the decision is to be made.

4° - Checking Performances

Care must be taken to see that the performance required of the human operator are on a satisfactory level and remain there.

In the same way that the inspection service sees to the quality of equipment and its installation and operation, internal or external audits must be used to check that the recommendations described above are effectively applied and that there is no slow degeneration in plant organization.

HUMAN FACTORS CONSIDERATIONS IN THE SAFETY CASE FOR THE CHANNEL TUNNEL PROJECT

ANDREW J SMITH BSc (Hons)
WS Atkins Engineering Sciences,
Stanford House, Birchwood, Warrington, Cheshire, WA3 7BH, UK

ABSTRACT

This paper outlines how human factors considerations are being incorporated into the safety case for the Channel Tunnel Project. Consideration of human factors aspects is given to both normal and post-accident operations. It is shown how consideration of these aspects has implications for operational procedures, staff training, staffing levels, improvements to the design and requirements for further testing and research.

INTRODUCTION

When completed, the Channel Tunnel System will link the road and rail networks of the UK and France. It will improve communications and stimulate commerce within the European Community by providing a fast and reliable service for cross-Channel traffic. The system is scheduled to open in May 1993.

The British and French Governments have set up a Safety Authority whose responsibility it is to act on their behalf in the interests of the travelling public on all safety issues relating to the Channel Tunnel Project. Thus it is necessary for Eurotunnel, the owners and operators of the system, to produce a safety case which is satisfactory to the Safety Authority before the system can be brought into operation.

Due to the many novel concepts and features of the system, any safety case produced for the Channel Tunnel Project would be incomplete without a thorough investigation of the interaction of these features with the passengers travelling on the system and the human factors concepts involved.

SYSTEM DESCRIPTION

The Channel Tunnel is a rail based transportation system which will allow both special shuttle trains and BR/SNCF through-trains to cross the channel. The purpose-built shuttles will carry road vehicles and their passengers. Trains will travel through the tunnel a total of 50km with 38km undersea at a top speed of 160km/hr. Each journey will last around 30 minutes. During peak periods it is intended that shuttles carrying passenger vehicles will run approximately every 12 minutes with further shuttles carrying freight vehicles every 15 minutes, although the system will be capable of development to permit more frequent operation.

The basic design for the Channel Tunnel System has been developed so as to be inherently safe. However, as with any transportation system, there remains a residual risk and this must be analysed in order to further improve the design, develop procedures and to assist in the preparation of training requirements. Later on in the project, when the final design has been ascertained, a quantitative analysis will be undertaken to show that the residual risk is acceptable.

The system consists of three separate tunnels; two running tunnels for train operation and a service tunnel positioned between the running tunnels. By providing two separate running tunnels the risk of an incident on one line obstructing and causing a collision on the other line is virtually eliminated. Similarly a fire in one running tunnel should not affect the other.

The service tunnel fulfils four functions. It provides ventilation to the running tunnels; it provides safe refuge for passengers and staff following an incident; it acts as access for emergency services and maintenance personnel and also will carry some of the service pipes and cables. Walkways are located in the running tunnels for maintenance work and evacuation purposes. Cross passages link the running tunnels to the service tunnel every 375m along the tunnel. Two cross-overs allow sections of the tunnel to be closed for maintenance while operations continue in single-line mode.

PROJECT ORGANISATION

The basic instrument authorising and regulating the Channel Tunnel System is the treaty between the UK and France, which was ratified on 29th July, 1987. Eurotunnel has entered into a Concession Agreement with the UK and French Governments under which it has undertaken to develop, finance, construct and operate the system. The concession entered into force on 29th July 1987 for a period expiring on 28th July 2042.

Eurotunnel has entered into a contract with a joint venture of ten major companies, five British and five French, which is known as Trans-Manche Link (TML). The contract Eurotunnel has entered into with TML specifies that TML are responsible for the design and construction of the tunnel and its facilities, providing a fully-operational system by May 1993.

In addition, Eurotunnel has appointed an independent organisation to monitor the design, development and construction of the works and to advise on technical and other matters; this organisation is known as the Maitre d'Oeuvre (MdO).

SAFETY IN THE CHANNEL TUNNEL PROJECT

An Inter-Governmental Commission (IGC) has been set up by the UK and French Governments to supervise, on their behalf, all matters concerning the construction and operation of the Channel Tunnel Project. Within the IGC a

Safety Authority (SA) has been established to advise and assist the IGC on all matters concerning safety in the construction and operation of the Channel Tunnel Project. Eurotunnel's design and operations proposals are subject to the scrutiny and approval of the SA and will have to meet all their requirements before the system can be brought into operation.

The MdO advises the SA on the adequacy of Eurotunnel's submissions and arbitrates in cases of disagreement between Eurotunnel and the SA. Objection to Eurotunnel submissions may only be on grounds of safety, defence, security and environment or non-conformity with general characteristics of the Channel Tunnel Project as defined in Annex 1 of the Concession Agreement.

HUMAN FACTORS ASPECTS OF SAFETY

Clearly a key element in any safety submission for the Channel Tunnel is that of human factors, due both to the large number of passengers being transported through the system and the unique nature of the hazards that may be encountered.

In order to demonstrate how human factors considerations are being satisfactorily incorporated into the overall safety case for the Channel Tunnel Project, this paper outlines a study that has been carried out by the MdO which examines the safety of operating the system by means of the non-segregation principle.

INCORPORATION OF HUMAN FACTORS INTO THE SAFETY CASE
FOR THE NON-SEGREGATION PRINCIPLE

Eurotunnel are proposing to design and operate the Channel Tunnel System according to the non-segregation principle. This principle leads to a system whereby passengers remain with their vehicles in the same shuttle wagon during the journey.

It is thus necessary for Eurotunnel to demonstrate to the SA that the system, based on non-segregation, is acceptably safe in both its design and operation. The MdO is required under the Concession Agreement to provide the SA with an independent report on safety matters. It was therefore necessary for the MdO to carry out an assessment of the implications for safety of non-segregation.

As part of this assessment, consideration was given to the implications for human factors of operation by means of non-segregation. Human factors aspects were considered both for normal operations within the tunnel and following emergencies.

Normal Operations

The following aspects of normal operations have the potential to affect passengers travelling with their vehicles:

Ventilation requirements: The major loading on the ventilation system of the shuttle is posed by vehicle exhaust gases during loading/unloading operations, which could in the absence of forced ventilation lead to carbon monoxide build-up in the atmosphere of the shuttle. To combat this it is proposed to purge shuttle atmospheres during loading/unloading operations using platform based equipment. Operational procedures will ensure that passengers switch off their vehicle engines immediately on entering the shuttle and that they do not subsequently switch them on again until it is time to leave the shuttle.

Other HVAC requirements: Studies have shown that unacceptable temperature conditions would occur within a shuttle during summer/winter conditions. It is therefore proposed to install on-board HVAC equipment. Experiments have also been performed to investigate the temperature and relative humidity increases on an occupied double decker coach. With the coach fully sealed and its internal ventilation shut off, conditions became uncomfortable over the duration of a shuttle journey. However, if the coach is allowed natural ventilation or the internal ventilation is on, conditions will remain acceptable.

Ride conditions: It was necessary to investigate the ride characteristics of the shuttle to determine if the double action of the road vehicle springing on top of that of the rail vehicle is likely to have any detrimental effect on passenger comfort. Results of tests indicated that an exposure of at least four hours would be required before passenger discomfort occurs due to these vibrational effects. As a journey within the tunnel is projected to be of the order of 30 minutes this would appear acceptable.

Other comfort considerations: Normal internal wagon illumination is intended to be 300 lumens/m^2, with higher levels to be provided at those areas used by passengers, such as toilet facilities. When the standby electrical supply is in operation the minimum illumination will be 50 lumens/m^2. Noise levels within wagons will not exceed 80 dbA at the shuttle's maximum speed of 160 km/hr. This level of noise will not significantly affect passenger comfort. Windows will be provided along the sides of wagons, even though there will be little of interest to observe, in order to avoid any claustrophobic effect upon the passengers.

Anti-social passenger behaviour: A factor which must be considered is the potential distress to other passengers and vandalism to the shuttle caused by anti-social elements amongst the passengers. Careful thought is being given to the prevention of any such behaviour by passengers on the shuttle, by such means as the refusal of admittance to undesirable groups, no alcohol to be on sale at terminals and the establishment of transport police stations at terminals. Extra vigilance will be employed where trouble is anticipated, such as the day prior to international football matches.

Post Accident Operations

Consideration must be given to human factors aspects in the highly unlikely event that an accident occurs within the Channel Tunnel System. The following aspects of post-accident operations within the tunnel have the potential to affect passengers travelling with their vehicles:

Hazards due to fire: The major hazards presented in post-accident operations are those which will occur during the outbreak of fire inside a wagon. The safety of occupants in a wagon in the case of fire may be threatened by a number of factors such as high temperatures, dense smoke, oxygen depletion, increase in carbon monoxide and carbon dioxide levels and the evolution of noxious products in combustion gases.

The consequential effects of these factors upon wagon occupants are different. High air temperatures can cause damage to the respiratory tract. Low oxygen levels and high carbon dioxide levels will cause asphyxia. High carbon monoxide levels and the evolution of noxious combustion gases can have toxic effects. Dense smoke will cause difficulties with visibility, spatial orientation and in addition further difficulties such as eye irritation. An assessment of these effects is a feature of an ongoing programme of fire tests using a mock-up of a shuttle wagon.

Hazards associated with fire extinguishing agents: The fire extinguishing system which is envisaged for the shuttle wagons will be of the total flooding type, in which the whole wagon would be bathed in a mixture of Halon 1301 and air, with a concentration of Halon 1301 at around 5%. The fire extinguishing system may be activated before all passengers have been evacuated from the unaffected wagon, or may spuriously activate under normal conditions, and therefore human exposure to Halon 1301 must be investigated.

Toxic hazards associated with Halon fire extinguishing agents are described in [1]. Halons act on the human body primarily as anaesthetics. Anaesthesia has no lasting effects on the human body. Halon 1301 is reported as leading to some lung damage but only after long-term exposure, which would clearly not be the case here. Halons are also thought to cause an increased sensitivity of the heart to adrenalin. This effect will vary between individuals. Clearly, those with heart defects are most at risk.

These effects are generally thought negligible at Halon 1301 concentrations below 10%[2]. As the design concentration is currently intended to be of the order of 5%, halon exposure should have negligible consequences. However, further work is being undertaken to confirm that this is the case.

The potential for Halon 1301 breakdown, in the course of extinguishing a fire, leading to the creation of dangerous decomposition products must also be considered. Decomposition products of Halon 1301 due to fire are mainly hydrogen fluoride (HF) and hydrogen bromide (HB) with possibly small amounts of bromine (Br_2). In addition there may be very small amounts of carbonyl fluoride (COF_2) and carbonyl bromide (COBr2) although this is not certain. Studies have shown, however, that decomposition of Halon 1301 does not lead to levels of these products which exceed their recommended short-term exposures[3].

The implications of a spurious release of Halon 1301 into a wagon have also to be considered. Halon discharge will result in loud noise levels, rapidly reducing wagon temperatures (by 10°C) and timbre of human voice changes. In order to reduce adverse passenger reaction, prompt information will be provided as to what may occur and what actions to take in the event of a spurious Halon release.

Vehicle movement following normal or emergency braking: Tests have been conducted in order to verify that road vehicles will not move either forwards, sideways or backwards when normal or emergency braking is applied to a shuttle.

There are a number of potential hazards associated with vehicle movement within the shuttle. Firstly, during normal operation a passenger or member of Eurotunnel staff could be trapped between two vehicles, leading directly to injury. There is also the hazard of entrapment between two cars occurring following an order to evacuate the wagon leading both to direct injury and also the failure of the trapped person to be able to evacuate the wagon. In addition, damage could occur to the pass doors at the end of the wagon, or to the lateral wagon escape doors, or blockage could occur of the evacuation pathway within the wagon.

The tests have shown that providing cars are secured by the handbrake and parked in first gear then no movement during normal or emergency braking will occur. However, small movements of the vehicles occurred with a reduction in handbrake efficiency and/or reduction in gear resistance. Following these tests it is now intended that the first and last cars within a wagon shall be secured. Further tests are envisaged to investigate movement of other types of vehicle.

There are several aspects of human factors within the non-segregation concept which in this area will provide a positive benefit for safety. Firstly, the driver can take immediate remedial action on detection that his vehicle is moving due to failure to apply the handbrake and/or park the car in gear. It is also the case that it would be a violation of a populational stereotype not to engage the handbrake upon parking the car and stopping the engine (in other words the driver expects to engage the handbrake under such circumstances).

This means that the probability of a driver failing to engage the handbrake is likely to be small. In addition, in the event of a major rail accident, the passengers are protected by a reasonably crash proof box which will afford a degree of protection well in excess of that normally provided on high speed trains. This is particularly the case if the passengers keep their seat belts fastened.

Vehicle evacuation: In the event of an incident in a shuttle wagon requiring evacuation, passengers will be directed to adjacent wagons, the affected wagon will be sealed and the shuttle will continue on to its destination where passengers will disembark. If for some reason the shuttle cannot continue out of the tunnel, passengers will be evacuated onto the walkway in the tunnel and from there will be directed to the service tunnel via a cross-passage. The passengers will then be evacuated from the service tunnel by train in the other running tunnel or by the service tunnel transport system.

Tests have been carried out to study human behaviour under simulated conditions requiring evacuation, primarily to determine the time taken for passengers to leave the affected wagon. The tests were carried out using cars and coaches in a full-scale dimensionally accurate shuttle wagon mock up.

The data for the simulations were analysed and interpreted in terms of three clear phases that have been identified in the evacuation process[4].

The first phase is _recognition_ which refers to the realisation by the passengers that something unusual is happening that may require special attention and behaviour. The second phase is _decision_ which refers to the taking of appropriate action in response to the cues received from the recognition phase. The third phase is _action_ which refers to the evacuation from the wagon, or taking some other action to deal with the situation as perceived from the decision phase. The minimum and maximum times for each phase for car occupants are shown on Table 1.

TABLE 1

Response time for each phase in the evacuation process

Phase	Minimum Response Time	Maximum Response Time
Recognition	0 seconds	in excess of 2 minutes
Decision	1 second	1 minute 17 seconds
Action	8 seconds	1 minute 20 seconds
TOTAL	9 seconds	in excess of 5 minutes

It can be seen from Table 1 that there is a wide variation in evacuation times. If all passengers were to take the minimum possible time in evacuation then the time for everyone to evacuate the wagon would be around 20 seconds. It is possible for the evacuation to take in excess of 5 minutes however, with the majority of this time being due to the recognition phase. This is to be expected as in situations of ambiguity people may ignore signals for a considerable time. Thus unambiguous cues will be provided to passengers in order to minimise evacuation times.

The simulations therefore show that although effective design for rapid egress is essential, the development of a reliable and effective information and alarm system is of equal importance. Both are technically feasible.

If people respond reasonably to appropriate warnings it is still possible for an evacuation of a wagon to take at most 1 minute 17 seconds of decision time and then a further 1 minute 20 seconds in time to act effectively. A reasonable time to keep the compartment in a safe condition during an emergency would appear to be around 3 minutes. Around half this time, 1 minute 45 seconds, made up of 15 seconds recognition time, 30 seconds decision time and 1 minute action time, could be regarded as typical of a situation in which the passengers were unaware of likely dangers, but well informed of what to do in an emergency.

It is likely that a real emergency could well have longer recognition and decision times, but the action time is likely to be reduced. Therefore, the overall typical evacuation time seems a reasonable first estimate at the present.

When compared with the results from fire tests described above, this typical evacuation time is generally adequate in order to evacuate all passengers before conditions become intolerable, although further fire tests are planned to more accurately determine the available time for evacuation.

Panic behaviour: Consideration has been given to the likely behaviour of the passengers when faced with a real emergency, in order to be able to determine the validity of the results of the evacuation trials described above.

The word 'panic' is frequently misused in describing the behaviour of people fleeing from danger. In many cases fleeing is the only sensible course of action, the critical difference between rational and panic behaviour being in the manner in which escape is effected. A useful definition of panic is: fear induced behaviour which is non-rational, non-adaptive and non-social and which serves to reduce the escape possibilities of the group as a whole[5].

A review of the psychological literature, e.g. [6]-[8], indicates that true panic is in fact a rare phenomenon, and given adequate guidance and information, evacuation is likely to take place in an orderly manner. Avoidance of panic during evacuation will be achieved by:

(i) provision of sufficient staff on each shuttle who are accepted by the passengers to be in charge and who are seen to act decisively for the benefit of the group as a whole, and,

(ii) provision of an effective alarm and information system which will both warn of danger and instruct as to the correct course of action. All passengers on embarkation will receive a hand-out outlining the actions that may be necessary in the event of an emergency. As much as possible of this information will be represented pictorially in order to avoid potential language difficulties. Information sheets will be printed in different languages, with the relevant sheet being given to each vehicle at the toll booth, according to the nationality of the passengers.

Other non-adaptive behaviour: The opposite reaction to panic in an emergency must also be considered, that of passengers underestimating the danger and remaining in their vehicles when told to evacuate. An alternative reason for passengers remaining in their cars could be a feeling of security within the car when conditions outside look threatening. In addition feelings of confusion and uncertainty can lead to delays in evacuation commencement. Again there is a great responsibility on the shuttle staff to ensure that all passengers have evacuated and that sufficient information has been passed on for them to fully appreciate the emergency.

Disabled passengers: The potential impact upon the safety of disabled people is recognised and further work is being carried out in this area. This work is complicated due to the large number and varying degrees of different incapacities that need to be considered, including physical disabilities such as blindness, deafness or loss or use of limbs, etc, medical conditions such as asthma or weak heart, and in addition many different types of mental disability.

CONCLUSIONS

This paper has demonstrated how consideration of human factors aspects is contributing to the safe design and operation of the Channel Tunnel System.

The work outlined in this paper has only looked at one aspect of the system, that of non-segregation, but it can be seen how the same human factors principles will be applied to other areas of the Channel Tunnel in order to ensure as safe and efficient a system as possible.

REFERENCES

1. Thorne, P.F., The toxicity of Halon extinguishing agents. Fire Research Note No. 1073, Fire Research Station, Borehamwood, 1977.

2. Reinhardt, C.F., and Reinke, R.E., Toxicology of halogenated fire extinguishing agents. Proceedings of symposium, National Academy of Sciences, Washington, USA, 1972, pp. 67-78.

3. Ford, C.L., Extinguishment of surface and deep seated fires with Halon 1301. Proceedings of symposium, National Academy of Sciences, Washington, USA, 1972, pp. 158-172.

4. Tong, D. and Canter, D., The decision to evacuate: A study of the motivations which contribute to evacuation in the event of fire. J. Fire Safe., 1985, 9, pp. 257-265.

5. Wood, P.G., The behaviour of people in fires. Fire Research Note No. 953, Fire Research Station, Borehamwood, 1972.

6. Canter, D., Studies of human behaviour in fire, Building Research Establishment Report, Fire Research Station, Borehamwood, 1985.

7. Quarantelli, E.L., The nature and conditions of panic. Am. J. Sociol., 1954, 60, pp. 267-275.

8. Sime, J.D., The concept of panic. In Fires and Human Behaviour, ed., D. Canter, John Wiley and sons. 1980.

23

"SAFETY AND HUMAN FACTORS IN MANNED SPACE FLIGHT SYSTEMS"

Ian Jenkins, BSc.
Messerschmitt-Bölkow-Blohm GmbH
Space Communications and Propulsion Division,
Munich, Federal Republic of Germany

ABSTRACT

This paper summarises the NASA Safety review process for Space Shuttle Payloads, showing trends and changes which have occured since the Challenger accident. Technical safety requirements for different programs are presented. The Safety Program for the European manned space programs are outlined.

1. INTRODUCTION

The Space Shuttle Challenger accident and the follow up investigations have been well reported and have received much public attention. Less well known are the management procedures and technical requirements imposed to assure that payloads using the Space Transportation Systems (STS) are indeed safe for this system and its crew. The paper reviews these, and highlights trends and changes arising as as result of the findings of the Accident Investigation Commission. The paper will make much use of practical experience with previous and current STS payloads, and will evaluate management and decision making processes related to safety and human factors.

High level decisions have recently been made to allow Europe to proceed with significant manned space activities, such as COLUMBUS, the HERMES

SPACE PLANE, and ARIANE V LAUNCH VEHICLE. The development of a safety programme is being performed by The Europeean Space Agency(ESA),the national space agencies(for example the French National Space Agency CNES), and by industry. The status of these activities will be presented.

2.The Safety Review Process for STS Payloads

From its conception, the STS was designed for a large variety of payloads coming from many different countries, organisations, and companies of wide ranging capabilities. Very early on, NASA management recognized the need to verify that STS payloads are safe for use in the shuttle, whilst at the same time not necessarly being involved in the technical/scientific goals of the payload or its mission success.

The NASA Johnson Space Center (JSC) document NSTS 13830 was developed to assist customers (Payload Organisations) of the STS in complying with system safety requirements established by NASA headquarters. The document explains in detail the safety analysis and assessment reviews. Figure 1 illustrates the responsiblities for implementation of Safety Requirements.

NASA headquarters has assigned the responsiblity to interface with the payload organisation and to review the payloads for adequate safety implementation to the STS Flight Operator at JSC and the STS launch/ landing site operator at KSC(KENNEDY SPACE CENTER).

There is a joint responsibility between NASA and the Payload Organisation for safety implementation. The Johnson Space Center (JSC) has overall responsibility for assuring the Safety of the STS for each mission, including interfaces between orbiter and payload and between different payloads. Prior to each mission, the JSC perform a mission level safety assessment.

The Payload Organisation has the responsibility for assuring the Safety of its Payload, as well as the effect of orbiter interface failure modes on his payload. The Payload Organisation has to perform an integrated Safety Analysis for the Payload,which includes the actual hardware in the STS and Ground Support Equipment(GSE) as well as Ground Operations at KSC.

Figure 1: NSTS Integrated Safety Program Responsibility

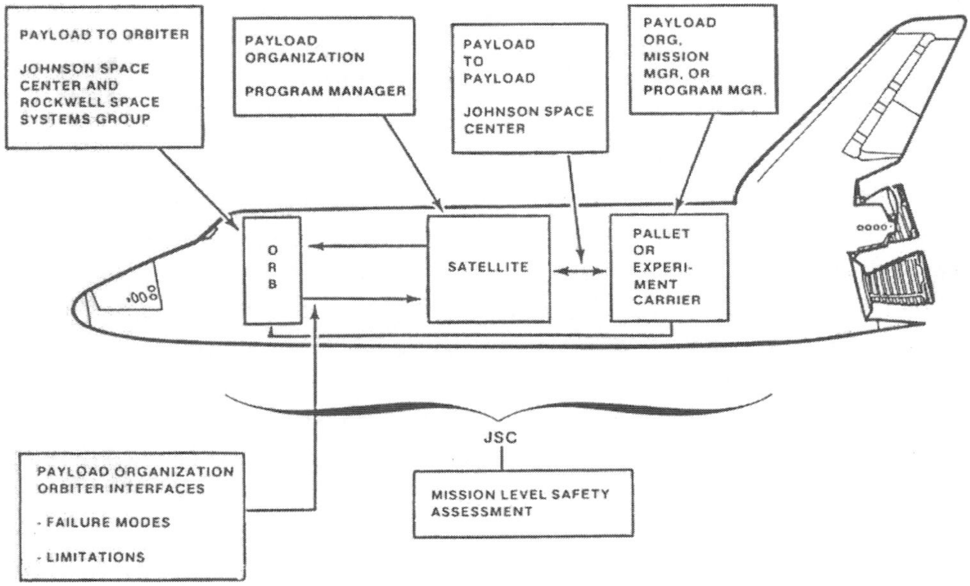

A series of reviews (Fig. 2) has been established to assure compliance with safety requirements.These reviews are intended to assist the payload organisations in interpreting the safety requirements, evaluate the payload safety and negotiate resulting payload safety issues.

Fig. 2:   Hazard Control Procedure

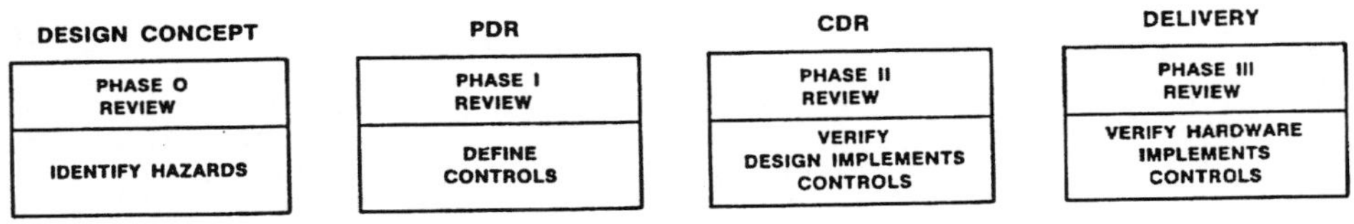

Note: PDR = Preliminary Design Review ; CDR = Critical Design Review

The phased series of four reviews is generally performed separately at JSC for flight and KSC for GSE/Ground Operations, because there are usually unique hazards associated with these operations. These reviews are linked into the payload development, starting at the conceptual phase through Preliminary Design Review(PDR),and Critical Design Review (CDR) to delivery.The review cycle is conceived to ensure that safety concerns are identified early during the design phase and that specific safety requirements can be implemented with minimum effort.

Table 1 shows the members of the Safety Review Panel.

The Flight Panel is chaired by the Payload Integration Office at JSC or the Systems Engineering Office at KSC. Members are drawn from all necessary disciplines at the NASA Centers. It is thus ensured that Panel decisions are shared responsibility, and not just of a specialist department. The role of NASA Safety Department is to disseminate safety data packages, coordinate responses, ensure correct representation at the safety reviews, ensure that all safety issues are addressed and resolved, and ensure that safety policy is consistant for different payloads.

Table 1 NSTS Payload Safety Review

Organisation		JSC		KSC
Function	o	Payload Design and Flight Operations	o	GSE and Ground processing
Chairman	o	JSC Payload Operations and Integration Office	o	KSC Systems Engineering Office
Members	o	Orbiter Integration	o	Ground Processing
Disciplines	o	Operations	o	Fuel/Cyro Handling
	o	Safety	o	Biomedical
	o	Engineering	o	Range Safety
	o	Ground Operations	o	Cargo Installation
	o	Life Sciences	o	GSE

Performance safety activities at a Payload Suppliers is dependant on the organisational structure. In order to allow flexibility, NASA have neither prescribed a specific organisational form nor required that safety specialists perform this work, however this may be required by American Deparment of Defence contracts. Thus it is possible for a small scientific institute to perform safety activities as an integrated part of the project. On the other hand, large aerospace companies such as Messerschmitt-Bölkow-Blohm(MBB) can support specialist departments because of the larger volume of work. At MBB, Safety Assurance is conducted under the umbrella of Design Assurance (see Figure 3) which is part of overall Product Assurance discipline.

Figure 3   MBB Project Safety Assurance Organigram

(Example: HERMES Project)

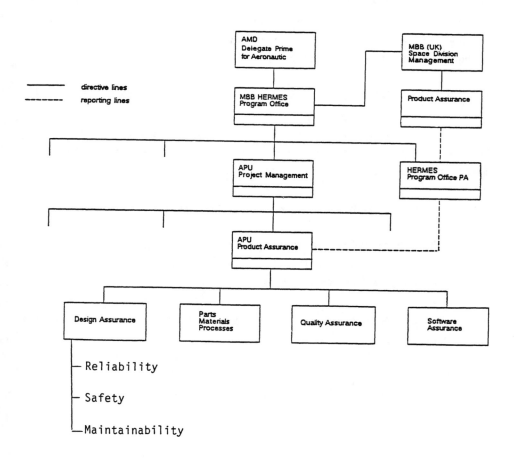

Whilst maintaining independance through separate reporting lines to the management, Safety Assurance specialists are assigned to specified projects. The specialists are familiar with NASA Safety requirements and can efficiently perform safety analysis. The safety specialist interfaces directly with all Project engineering and management staff. One of the most important tasks of Safety Assurance is to ensure that Safety requirements are correctly understood and implemented by engineering. Overall responsibility for Product Safety must remain with the project.

One of the strengths of the Safety implementation procedure is the relatively cost effective manner in which data are presented and reviewed. However, the strength of the system is also its weakness. NASA were not able, in all cases in the past, to adequately ascertain that Hazard Verification was being implemented. The implementation procedure and Safety Policy have been extensively revised as a result of the Challenger mishap to correct these deficiencies. The expected changes are outlined in Table 2.

3.   Technical Safety Requirements for Manned Space Flight

This paragraph summarizes some of the changes to the STS Payload Safety requirements and provides an overview of the current European Safety requirements for manned space flights.

Typical Safety Hazards for the STS are illustrated in Figure 4.

All of these hazards can be considered related to human factors in the sense that a hazard to STS is automatically a hazard to its crew. More immediate human factors result from; for example:

-   shuttle crew performing operations with hazard potential (Payload release etc.)
-   Offgasing and toxicity problems within "closed loop" inhabited areas
-   EVA operations, with astronaut contact to payload, etc.

Table 2   Major Changes to NASA Payloads Safety Implementation

Requirement	Background
o The NSTS Assessment of Safety Compliance will include a complete review of the Safety Assessment Reports and may include audits and inspections.	o NSTS always had the right to conduct audits and inspections but this was not explicetely stated o NASA performs these where concerns or noncompliances were identified.
o Minimum Supporting Data Requirements for "Design for Minimum Risk" Areas are defined in NSTS 13830.	o Minimum Data Set was not previously defined in formalised manner o Data submitted were not consistent toward establishing STS confidence in payload safety
o When manufacturing procedures/processess are critical steps in controlling a hazard, they shall be independantly verified in real time, and identified on the Hazard Report.	o NSTS return to flight activity increased emphasis on payload verification that safety requirements have been met.
o Phase III Safety Reviews shall be completed 30 days prior to ground processing.	o NHB1700.7A did not address completion of Flight Safety Panel Activities prior to ground processing o Ground Hazards are sometimes controlled by Flight Hardware Hazard Controls
o Flight Certification Statement shall include a Safety Verification Tracking Log	o Tracking and documentation of open Safety Verification after Phase III Safety Review

Figure 4: Payloads Hazards to the STS

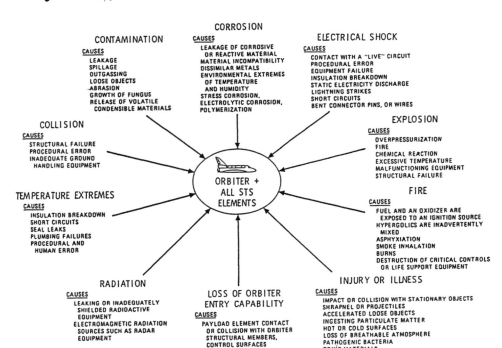

The NHB1700.7 (Ref 2) is the primary regulatory document specifying Safety Policy and requirements for payloads using the STS. This document has been extensively reviewed and revised since the Challenger accident, with nearly 130 detailed changes. Changes include new requirements resulting from the National Space Transportation System (NSTS) revalidation process updates to incorporate polices which were already being implemented, but not formally documented, and editorial changes for better clarity. The document has been extended to apply to man tended platforms as well as payloads carried in the STS. Man tended platforms are any space facility which can be serviced by the Space Shuttle Orbiter and can be provided by a "man rated" habitable area by the STS.

Table 3 shows hazard criticality classification for different NASA, European Space Agency (ESA) and French National Space Agency (CNES)

Table 3 Hazard Consequence Severity Categories

31

NASA NHB1700.7B
CATASTROPHIC HAZARD. A hazard which
can result in the potential for: a
disabling or fatal personnel injury;
or loss of the Orbiter, ground
facilities or STS equipment
CRITICAL HAZARD. A hazard which can
result in damage to STS equipment,
a nondisabling personnel injury, or
the use of unscheduled safing
procedures that affect operations of
the Orbiter or another payload

ESA PSS-01-40 Iss. 2, 17.03.88 (3)
I.   CATASTROPHIC
     -   loss of life, life threatening or occupational illnes
II. CRITICAL
     -   temporarily disabling but not life threatening injury, or
         temporary occupational illness
     -   loss of, or major damage to flight systems, major flight
         system elements, or ground facilities
     -   long term detrimental environmental effects
III. MARGINAL
     -   minor non-disabling injury or occupational illness
     -   minor damage to other hardware
     -   minor damage to public or private property, or
     -   temporary/detrimental environmental effects
IV. NEGLIGIBLE
     -   Will not result in any of the above

CNES H-SM-0-50-1, Iss. 1 (May 87) (4)
Management Spec. "Dependability"

"CATASTROPHIC" Loss of Human Life
lives
"SERIOUS" Loss of equipment or
serious injury
"MAJOR" Loss or non respect of
objectives
"MINOR" Minor degradation

CNES/AS AS.SM.0 50.1(1) (5)
Dependability (RAMS) ARIANE 5

0   "Fatal" Loss of human life
    and substantial destruction
1   "Major" Loss of mission,
    risk of injury to personnel,
    and limited destruction of
2   "MINOR" Controlled event
    with very limited effect,
    mission degraded
3   "NEGLIGIBLE" No noteworthy
    effect on execution of the
    mission (e.g. loss of redun-
    dancy, safety barrier, etc.)

AS/AMD H-SG-1-51-AS   (6)
RAMS-Studies HERMES

CATASTROPHIC: death (crew
or ground personnel)
HAZARDOUS: considerable
injuries, destruction of
spaceplane or ground facilities
MAJOR: mission interruption
MINOR: no or minor effects on
the spaceplane

Note: ESA and CNES Definitions are subject of a harmonization exercise during 1988

programmes. These classifications are important for determining fault tolerance requirements such as shown in Table 4.

Differences in definitions and terminology result from the different requirements of the programs. In particular, extensive ESA studies show that a reduced failure tolerance compared to NASA STS payload requirements is acceptable when the full ESA Product Assurance and Safety Requirements are applied to safety critical functions in manned space programmes such as Arianne V , Spacelab and Hermes. Currently ESA and CNES are harmonising their requirements. NASA STS fault tolerance requirements result from the consideration that a Payload Organisation may not necessarily be able to implement a full Product Assurance system as normally applied in space projects of high reliability.

One significant aspect of the European manned space programmes is the recognition for the need for crew escape and rescue, with special emphasis on the critical launch phase. MBB is performing definition studies on the safety critical HERMES Crew Escape Module. These studies are due to be completed in 1990, and will provide a challenge for Safety Assurance during this period.

4.   DEVELOPMENT OF A SAFETY PROGRAMME FOR COLUMBUS AND HERMES

The European Space Agency, the National Space Agencies, and industry are in the process of revising their Safety Assurance programmes, in order to be able to respond to the demands of the increase in manned space activities. MBB as a major space contractor has significant responsibilities within the Columbus and Hermes space programmes, and, as such, is involved in the review, definition and implementation of these safety policies.

The ESA Product Assurance(PA) and Safety Programme is aimed at a fully integrated European System for space application and close coordination with other fields of application.

A basic ESA Safety Policy is defined, which has the purpose to protect:
-      Human Life

33

Table 4   Failure Tolerance Requirements

NASA NHB1700.7B
STS PAYLOADS
Catastrophic Hazards. Catastrophic Hazards shall be controlled such that no combination of two failures or operator errors can result in the potential for a disabling or fatal personnel injury loss of the Orbiter,ground facilities,or STS equipment.

Critical Hazards. Critical hazards shall be controlled such that no single failure or operator error can result in damage to STS equipment, a nondisabling personnel injury, or the use of unscheduled safing procedures that affect operations of the Orbiter or another payload.

Mission success not regulated by NHB1700.7B

ESA PSS-01-40 Issue 2, 17.03.88 (3)

a) No single failure shall result in a catastrophic or critical
b) No single human operator error shall result in a catastrophic
c) No combination of a single failure and a human operational error shall result in a catastrophic hazardous consequence.

Mission success criteria in PSS-01-30

AS/AMD H-SG-1-51-AS 3.9.87 (4)
HERMES SPACEPLANE
FAIL OPERATIONAL   This criterion is used to prevent a single failure from causing a mission interuption. (Major Consequences)

FAIL SAFE   This criterion is used to prevent failures from leading to Catastrophic or Hazardous consequences:

- as a single failure(Fail safe)
- after a single failure which allows the mission to be continued(Fail Operational/ Fail Safe)

Probabilities of success are associated with these criteria.
Regulates Safety and Mission success

- Spaceflight Hardware, Ground Support Equipment and Facilities
- Public or private property, and
- the environment

from the hazardous consequences of ESA project activities.

The relation of ESA Safety Documents to the overall ESA PA System (PSS-01 Series)Requirements are shown in Figure 5.

Table 5 shows the status of ESA PSS-01-40 Safety Documents as of June 1988.

Table 5   ESA PSS-01-40-Status

Specification	Subject	Status
PSS-01-40	Safety Programme Requirements	Issue 2 (March '88) in industrial review cycle
PSS-01-400	Safety Data Package	Due Sept. 1988
PSS-01-401	Fracture Control	Issue 1 (March 87) already commented reissue mid 1988
PSS-01-402	Design Safety Handbook	Target: September 1988
PSS-01-403	Hazard Analysis	Target: September 1988
PSS-01-404	Risk Assessment	Target: September 1988
PSS-01-40?	Sneak Circuit Analysis	Under study contract

MBB and Aerospace Industry provide, via Eurospace, review comments to the ESA PSS System. MBB's comments are based on extensive experience with the STS Payloads, and have been benifical in providing practical comments on these specifications. Curently a major effort is under way to harmonize the programmes of the National Space Agencies with ESA requirements.

Implementation of Safety Policy is illustrated in Figure 6. This programme provides:
- a deterministic Safety Programme
- Hazard Reduction and Control

- Safety Reviews
- Progressive Risk Assessment, phased to project needs.

MBBs PA and Safety Program has been under rationalisation and is intended to be compliant with ESA requirements. By use of standardization MBB is able to cost effectively implement a safety program, whilst maintaining flexibility with respect to special needs of different projects such as NASA payloads, ARIANE V, HERMES and COLUMBUS, the organization allows for dissemination of safety lessons learned between the projects.

5.Conclusions

NASA have an efficient Safety Review Process for STS Payloads. Numerous changes have been made to the implementation procedure to improve closed loop accounting. The STS Payload Safety Requirements have been totally reviewed since the challenger accident. The changes present an improvement in understanding basic requirements, and an extension to include space station, and man tended platforms.

For the European manned space programs including HERMES, ARIANE, and COLUMBUS, ESA and european industry are developing safety policy and requirements emphasis must be placed in the near future, on harmonizing and standardising these requirements to assure a cost effective safety programme.

Acknowledgements
The author would like to acknowledge the support and encouragement of the MBB-Space Communications and Propulsion Systems Division in production this paper.

Figure 5: ESA PA System Hierachy

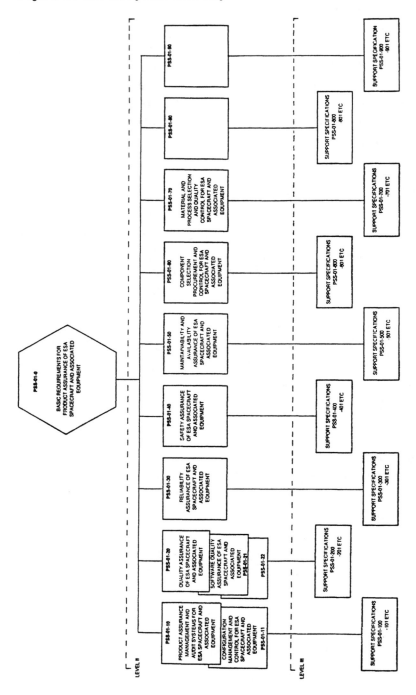

Figure 6: Safety Programme Life Cycle

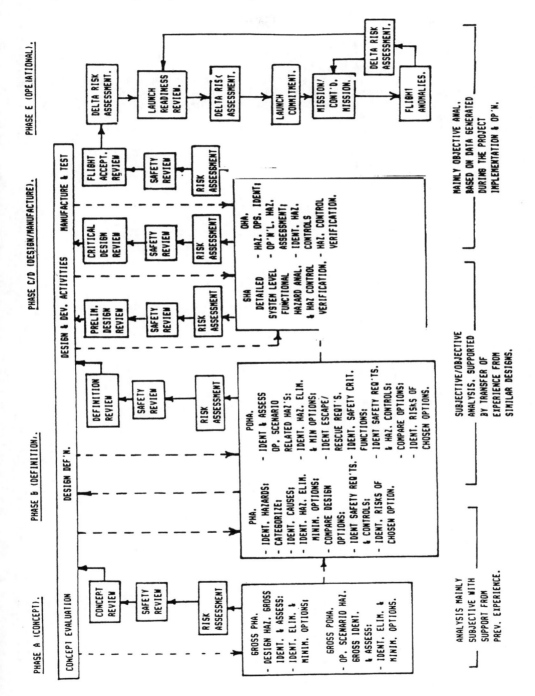

References

1. NSTS 13830   Implementation Procedure for STS Payloads System Safety Requirements. NASA-Lyndon B Johnson Space Center
2. NHB1700.7B   Safety Policy and Requirements for Payloads Using the STS. NASA JSC 2.2.88
3. PSS-01-40 Issue 2   System Safety Requirements for ESA Space Systems and Associated Equipment, ESA, 17.03.88
4. H-SM-0-50-1 Issue 1   Management Specification "Dependability", CNES, May 1987
5. A5.SM.0.50.1   Dependability (RAMS Reliability, Availability, Maintainability, Safety)
6. H-SG-1-51-AS   HERMES Program-Rules for Safety and Reliability, Aerospatiale

THE ROLE OF TECHNICAL DOCUMENTATION FOR MAINTENANCE STAFF IN THE SAFE OPERATION AND MAINTENANCE OF MACHINE AND PROCESSES

DR K CLEMENTS-JEWERY
(ASSIST/HEAD – SCHOOL OF ENGINEERING GLASGOW COLLEGE)

ABSTRACT

This paper explores the existing use (and mis-use) of technical documentation and drawings by staff engaged in maintenance - fault diagnostic activities on complex and potentially hazardous machines and processes. Explanations are put forward as to why many employees ignore or fail to understand the underlying logic in the use of information which in some instances has resulted in death or injury to people. This information has been drawn from the personal experiences gained from in-company technical training and consultancy across a very wide spectrum of UK industry. Suggestions are made for improvement in this man-machine-information interaction with resulting benefits of improved efficiency in the maintenance activity ie reduced downtime and safer working environment for all staff coming into contact with ever more complex manufacturing processes and systems.

1. THE ROLE OF TECHNICAL DOCUMENTATION IN FAULT DIAGNOSIS

Fault diagnosis in a system is greaty helped by well prepared specially written documentation which is used to prompt and guide the diagnostician to identify the original cause of the fault.

In order to carry out the task of fault diagnosis three prerequisites are necessary:-

i) <u>Adequate written details and drawings</u>
Providing enough detail to carry out the task. Irrelevant details such as used during manufacture of the machine but which is not necessary for the task is removed from the documentation.

ii) <u>Ability to understand these details</u>
The diagnostician needs to be trained in the technical subject areas that are intrinsically present within the machine or system. Eg. CETOP symbols used in pneumatics needs to be understood by the diagnostician in order to make sense of the pneumatic circuit diagram.

iii) <u>The capability of the diagnostician to logically locate the fault</u>
This requires the use of a logical sequence of elimination of possible causes of the symptoms currently observed. This task normally requires the diagnostician to be thoroughly familiar with the normal fault-free operation of the machine prior to breakdown.

iv) If any of these key requirements (i-iii) is missing then the fault diagnosis task degenerates into hit and miss servicing with consequent increased production downtime and less safe operation for the diagnostician.

Basic underlying principles in technical documentation for fault diagnosis

In most control systems a chain of events starting from an initial stimulus eg a start switch and terminating in the operation of powered device eg. an electric motor of hydraulic activator can be identified. In between the start and ending of the signal path many intermediate steps are normally present. A left to right flow of signals either "mechanical" or "electronic" passes through the chain as illustrated:-

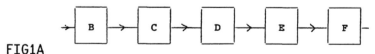
FIG1A

In many cases feedback signals are used (as in closed loop system):-

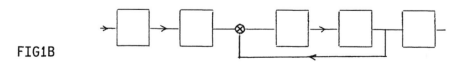
FIG1B

The main task in fault diagnosis is to identify the failed item within this chain in the shortest time with the least amount of testing. To this end "split half" techniques are quite useful. To illustrate this method assume that the final operation of the box F in FIG1A does not operate due to a fault in the chain. Measurement taken at the ouput of Box C will indicate if correct that the faulty item must be located downstream of C ie. to the right of C. This now means that half the potentially faulty items have now been eliminated. If now the half-split is applied again only 25% of the problem still exists, 75% of the potentially faulty items in the chain have been given a clean bill of health. Thus fault diagnosis relies on logical elimination of potentially failed items and therefore all technical documentation should if possible be based on this principle being applied.

2. CREATION OF DOCUMENTATION

Modern manufacturing and process plants contain a very wide and ever increasing range of mechanical, electronic and instrumentation systems which are designed by a number of specialist engineers. These engineers are required to provide source material for various types of documentation (including that for maintenance purposes) in a form suitable for the end user.

Consider as an example the case of an NC Machine tool product - a now very common and familiar item of production equipment in many factories. To design and maintain this equipment a team consisting of (at least) the following skilled staff is necessary:-

   i) A mechanical engineer.

  ii) An electronic/electrical engineer.

 iii) A manufacturing technologist.

  iv) A maintenance/service engineer.

These four staff have to communicate with each other during the design and construction phase and in the fullness of time with the eventual customer (owner) of the machine.

Documentation in the form of drawings, instruction, parts lists, test schedules etc have to be used as the main method of transference of vital information to the end user. Through daily interaction, this team - consisting mostly of peers of the same professional level - can ensure without too much difficulty, that the transfer of vital details and data one to another is efficiently carried out. Transfer of this information to the very wide spectrum of education and experience levels found with the end user is not so easy.

3. TRANSFER OF THIS INFORMATION TO THE END USER

The information amassed by this design team is often compiled into some form of equipment handbook comprising text and sets of drawings which is then delivered with the machine to the end user.

From my own experience in industry (where this information was needed for efficient maintenance) I have found that considerable problems occur in the quality of the material and the degree with which the customer understands what he has been given. There is often a very low efficiency in the process of transferring the information from the machine designer to the machine user and this can result in hazardous or even fatal consequences to the staff who operate or maintain these machines. The underlying reasons why this should happen are worthy of analysis and hopefully from the investigation, measures can be set up to reduce these hazards to the user.

4. **SOME EXPLANATIONS AS TO WHY THERE IS THIS LOW EFFICIENCY OF INFORMATION TRANSFER**

(a) Mismatch in education and experiences.
A considerable difference in educational attainment and experience often exists between the originator (designer) of a machine and the user of this machine. eg A graduate electronic engineer may compile a list of test sequences and instructions (which he understands) which are not properly understood by say a maintenance electrician who holds perhaps a City and Guilds trade certificate.

(b) Depth and breadth problem.
The designer and the manufacturer of a machine or system has of necessity a very deep, expert knowledge of that particular machine or system. The user however has normally a whole range of different machines and processes to maintain within the factory and thus is not very likely to have same depth of knowledge as the designer.

Thus either the designer has to compile material specially written to be understood by the user (a technical authorship skill) or the educational and skill levels of the user must be raised by training courses. Both actions may be undertaken which should result in a much closer match between the technical author and user thus increasing the efficiency of information transfer.

(c) Presentation and form of documentation.
The understanding and acceptability of documentation or instruction is greatly influenced by the style of presentation. The use of algorithms for instance can greatly clarify the thinking process in the unravelling of information from text. For instance consider the written instruction issued in leaflet N195 shown in Appendix I page (i) and compare this with the algorithm shown in page (ii). A significant clarification for the reader is evident - who can now understand the message which may have previously caused confusion.

Appendix II pages (i)-(iii) illustrate a programmable logic controller (PLC) process plant which operates to a sequence of emptying various tanks, heating and pumping into a storage tank. Page (i) shows the plant layout page (ii) shows the ladder program. code which is fed into the PLC to control the plant. This information is (at best) all that is normally provided by the designer or manufacturer of the plant and would be available within a handbook or set of drawings. However in the event of a fault developing in a system the fault diagnostician needs to be able to read and interpret the ladder diagram and code in order to diagnose and repair the fault in the quickest and safest way possible. If however the information in page (iii) in Appendix II is provided, showing the operation of the plant in the form of LED display and phase diagram the diagnostician will make a more valid analysis of the cause of failure. Without this additional aid - in the form of visual indication at each step of the cycle - many maintenance staff carry out trial and error experiments in the hope of locating the cause by chance. This compromises safe operation and should be avoided at all costs.

(d) Human inertia and procrastination.
Many people have a tendency to leave things alone until there is a problem eg one is only interested in maintenance when the car fails to start! However for good preventive and quick response breakdown maintenance there is a need to consult the documentation provided prior to breakdown or signs of trouble. Staff who wait until there is a crisis have then got to accumulate information from the documentation under duress - with attendant loss of efficiency and regard to their safety.

5. **RESULTS OF LOW EFFICIENCY OF INFORMATION TRANSFER - DISCUSSED UNDER 4 (a)-(d) ABOVE**

If one does not understand one ignores. The typical response to points discussed above is to ignore the problems - mostly by failing to use the documentation provided. From my own experience in training maintenance staff in over a 100 companies throughout the UK the phase "if all else fails read the handbook" seems to be almost a norm. This quite often results in damage to equipment, personal injury (near misses by the thousand) and very occasionally death because operations were carried out incorrectly through ignorance.

Ignorance is not bliss!

In addition there are resulting effects on downtime costs, employee frustration and low job satisfaction.

6. **WHAT CAN BE DONE TO IMPROVE MATTERS**

In the light of the foregoing comments measures can be put in hand to improve matters:-

(a) Presentation of material - already discussed. Use every aid possible.

(b) Awareness of problem by the designer or manufacturer of machine to address the current level of education/skill levels of the end user - this requires much more co-ordination between the back room boys and the 'shop floor'.

(c) Testing of material on a trial basis prior to giving it out a a general issue.

(d) Education of both user shop floor staff and management to ensure they are motivated to want to use the documentation.

7. DISCUSSION ON ONGOING DEVELOPMENTS

(a) Computer Aided Learning (CAL)

The critical role of fault diagnosis and the consequences associated with safety and reliability has received considerable attention in recent years. A very good summary of fault finding skills, training and methods of presentation has been produced by the MSC (Ref 1). Considerable literature is quoted in the Ref which records many experiments in industry which have taken place. New developments in forms of information transfer other than by written documentation are currently under development - such as computer based learning or instruction (CBL-CBI). These techniques often use feedback from the trainee in the correct answering of questions set by the CAL program in order to advance to the next instruction.

This interactive method of learning promises to ensure that information needed to service or maintain equipment is indeed understood by the user rather than assumed by more conventional learning techniques.

(b) Built-in machine diagnostics

With the ever lowering cost of computer systems many machine producers build machine diagnostics - MIMIC panels and VDU screen displays which in essence contain help that previously would have been continued in written and pictorial form within handbooks. This promises to significantly improve the efficiency of fault diagnosis and raise the level of safety for the user.

Both CAL and use of built-in machine diagnostics do however have a shared drawback namely that the maintenance staff have to not only master the machine but also have to learn to operate the computer in the case of CAL or the diagnostic system within the machine. This may cause problems for some staff in that if currently, they do not use handbook and drawings why should they use a CAL in order to understand the operation of a machine?. This is basically a problem for the management of the company who in many instances are unaware of the importance of good fault diagnosis documentation.

8. CONCLUSIONS

The key issue at stake is that of _motivation_ and _awareness_. If the employer of maintenance staff is unaware of the scale and importance of the problems discussed above then it is not surprising that the staff continue to muddle on. Very often senior management are unaware of the poor work situation and the workforce are sometimes aware of the problem but are not articulate enough to explain to the top management on a cost/benefit basis the advantages to be obtained by the improvement in documentation and system of working.

The costs incurred in setting up adequate documentation and in training staff to correctly use it can be substantial.

However, the hidden costs of inefficiency which is present in many manufacturing plants is considerably higher in real terms. Many companies have maintenance budgets in the order of £2M – £3M PA and based on the very modest overall improvement of say 10% on this budget which could be experienced by raising the standards as suggested in this paper the company could easily pay for the setting up costs, probably over one year.

Quantification of costs saved by prevention of claims raised by injured staff as the result of an accident are impossible to estimate but could be very substantial. If adoption of measures suggested in this paper are taken up the company can claim that all reasonable measure save been taken to prevent death or injury to its staff thus limiting the amount of the claim.

REFERENCE: MSc Publication No.26 by Patrick Barwell, Toms & Duncan.

Department of Applied Psychology, UWIST

46

APPENDIX I

(i)

ALGORITHMS

(Reproduced from "Design of Instruction" Dr Sheila Jones, D.E.P. Training Information Paper 1)

Extract from Ministry of Social Security Leaflet N.I.95(1967)

WOMEN WHOSE MARRIAGE ENDS BY DIVORCE WHEN THEY ARE 60 OR OVER

Benefits

12. If all other conditions are satisfied, you can qualify on your former husband's contribution record for a pension equal to that which you would have received had he died on the date your marriage ended, regardless of whether you have, or your former husband has, retired or not. If the marriage took place after you were 60 this rule will help you only if the marriage lasted three years or more or if, when you married, you were entitled to widow's benefit or retirement pension under the National Insurance Scheme or had to give up some other pension paid from public funds.

(ii)
ALGORITHMS

Ministry of Social Security Leaflet N.I.95(1967), para.12 as an Algorithm

(Reproduced from "Design of Instruction" Dr Sheila Jones, D.E.P. Training Information Paper 1)

RETIREMENT PENSION FOR WOMAN WHOSE MARRIAGE ENDS BY DIVORCE WHEN 60 OR OVER

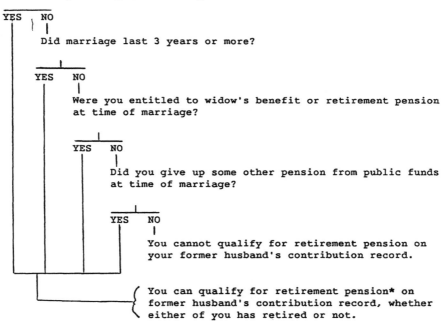

* Pension will be same as you would have received if husband had died on date marriage ended.

(i) APPENDIX II

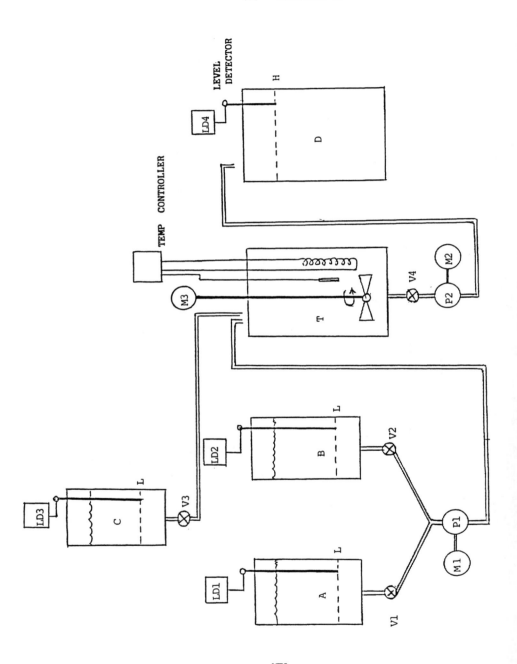

MEL
PLC PROCESS CONTROL RIG

(ii)

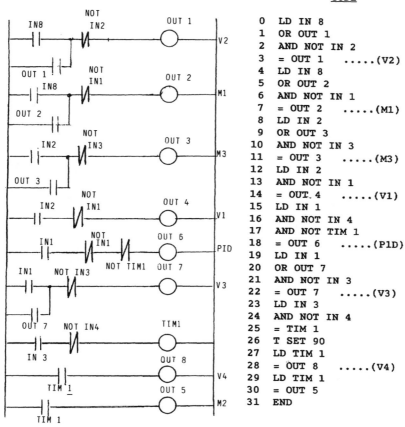

CODE

```
0   LD IN 8
1   OR OUT 1
2   AND NOT IN 2
3   = OUT 1     .....(V2)
4   LD IN 8
5   OR OUT 2
6   AND NOT IN 1
7   = OUT 2     .....(M1)
8   LD IN 2
9   OR OUT 3
10  AND NOT IN 3
11  = OUT 3     .....(M3)
12  LD IN 2
13  AND NOT IN 1
14  = OUT 4     .....(V1)
15  LD IN 1
16  AND NOT IN 4
17  AND NOT TIM 1
18  = OUT 6     .....(P1D)
19  LD IN 1
20  OR OUT 7
21  AND NOT IN 3
22  = OUT 7     .....(V3)
23  LD IN 3
24  AND NOT IN 4
25  = TIM 1
26  T SET 90
27  LD TIM 1
28  = OUT 8     .....(V4)
29  LD TIM 1
30  = OUT 5
31  END
```

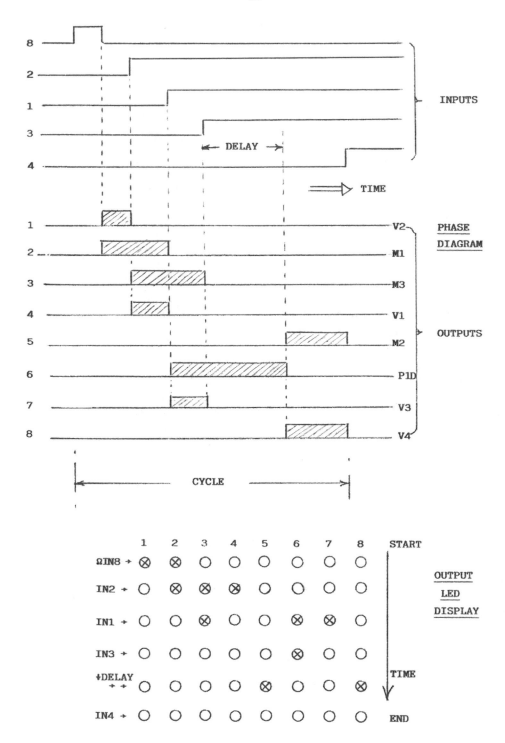

51

ALLOCATION OF FUNCTION BETWEEN MAN AND PROGRAMMABLE ELECTRONIC SYSTEMS IN SAFETY-RELATED APPLICATIONS

J BRAZENDALE
Control Systems Group
Technology Division
Health and Safety Executive
Magdalen House, Stanley Precinct, Bootle, Merseyside, L20 3QZ, UK

ABSTRACT

This paper considers the allocation of function between man and programmable electronic systems in safety-related applications. HSE's guidance on Programmable Electronic Systems is discussed and in particular the concept of a safety-related system explained. The relation of allocation of function to the **configuration system element** is amplified by way of an example.
The paper then discusses the role of allocation policy in determining system safety and how the concepts of inherent safety and human error of various types should form part of the allocation decision process.
Errors induced by personal and organisational factors are contrasted with cognitive errors and strategies for overcoming them given.
A number of problem areas are discussed and possible ways of taking the topic forward suggested.

INTRODUCTION

Feedback from Industry and others has indicated a demand for guidance on safety issues associated with Human Factors.

The Health and Safety Executive (HSE) has responded to that demand and in the near future proposes to issue a general guidance booklet on "SAFETY AND THE HUMAN FACTOR". This guidance is based upon the control strategy shown in Fig 1 which considers Human Factors in the context of Organisational factors; Personal factors; and Job design. The booklet is a first step to give guidance in this complex field and its aim is to stimulate managers to consider the whole range of human factors as a

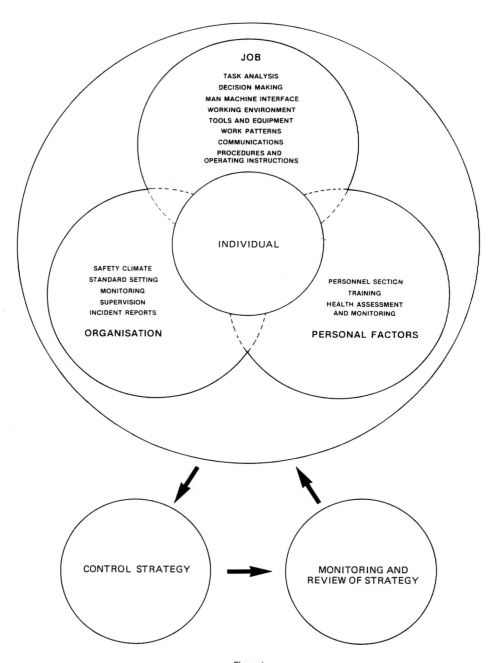

Figure 1
THE DEVELOPMENT OF A CONTROL STRATEGY
FOR HUMAN FACTOR FAILURES

distinct element which they must recognise, assess and manage if they are to control risk.

This general guidance will be supplemented by more specific guidance on particular topics.

Within the Control Systems Group of Technology Division of HSE the topic of allocation of function between man and computer-based systems is currently under review. This topic has been raised during feedback on HSE's publication on Programmable Electronic Systems (PES); and has also been found to be a factor in a number of incidents.

The purpose of this paper is to outline the PES strategy for safety; to show how the allocation of function problem is dealt with in the strategy; to highlight some problem areas; to suggest some ways forward and to obtain views and comments.

PROGRAMMABLE ELECTRONIC SYSTEMS

In the economic climate of today there is increasing pressure on Industry to automate the production process including many safety functions and safety features.

PESs have the potential to improve safety by, for example, providing sophisticated alarm analysis facilities. However, experience of incidents involving PES has shown that many of these advantages will not be realised unless a disciplined and structured approach to design is adopted at all project stages.

The PES is defined as a system based upon a computer connected to sensors and/or actuators on the plant for the purpose of control, protection or monitoring. The term includes all elements in the system extending from plant sensors or other input devices, via data highways or other communication paths, to the plant actuator or other output device.

That part of the PES which handles the logic processing is termed the "programmable electronics" and refers to those parts of the PES which are not solely dedicated to a particular sensor or actuator.

Fig 2 illustrates the basic PES structure.

In order to provide advice on a structured approach to this topic HSE published two documents in June 1987 which are the first in a series whose general title is "Programmable Electronic Systems in Safety-Related Applications".

The two documents are:

"An Introductory Guide" [1]

This document (PES1) is aimed at the non-specialist and provides an overview of the safety principles.

"General Technical Guidelines" [2]

This three part document (PES2) contains:-
(1) General guidance on the problems, and a framework within which they can be approached systematically.

(2) A method for assessing the safety integrity of PESs. (Outlined in Part 1 and described in detail in Part 2).

(3) A worked example using the method in Part 2 is described in Part 3.

SAFETY FRAMEWORK PRINCIPLES

Considerations Underlying Guidelines

The guidelines are generically based and have been structured so that they do not unreasonably constrain design innovation.

They are concerned with those PESs which either acting alone or in combination with non-programmable systems (including "human systems") provide the required level of safety. Such systems upon which the safety

FIGURE 2: BASIC PES STRUCTURE

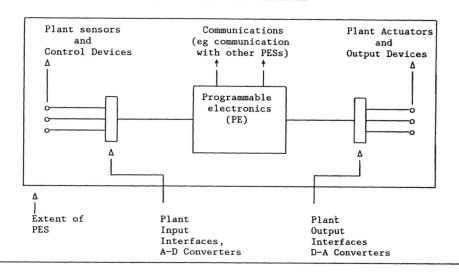

FIGURE 3: DESIGN AND ASSESSMENT STRATEGY

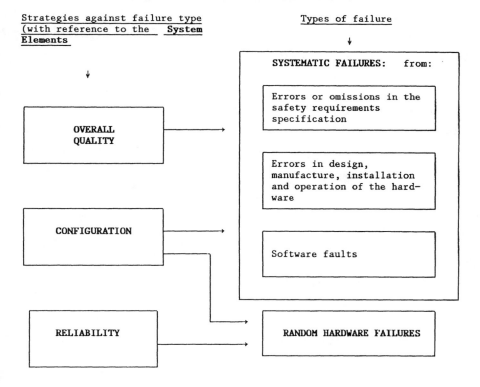

integrity of the plant relies are referred to as **safety-related systems.**

The key point as far as Human Factors is concerned is that the guidelines recommend that ALL safety-related systems are assessed as part of the process of designing a PES-based safety related system.

Safety Principles

The safety strategy underlying the recommendations made in the guidelines is centred on three system characteristics or **system elements.** The principles which govern the system elements underlie the design and assessment strategy for a safety-related PES. The three system elements are defined as follows:-

Configuration: The specific arrangement of the programmable electronics within a PES and the combination of PES and non-PES safety-related systems.

Reliability: That aspect of the safety integrity relating to random hardware failures in a dangerous mode of failure of the safety-related systems.

Overall Quality: The non-quantifiable qualitative aspects of the safety integrity of the safety-related systems. This system element is concerned with the precautions taken against systematic failures.

The detailed requirements of the three system elements are, together, intended to tackle both random hardware failures and systematic failures so that a defined level of safety integrity is achieved. (See Fig 3).

Random hardware failures are those failures which result from a variety of degradation mechanisms in the hardware. Measures of reliability such as the "mean time between failures" are concerned only with random hardware failures and do not include systematic failures.

Systematic failures are concerned with errors in the design, construction or use of a system which cause it to fail under some specific combination of inputs or under some environmental condition. Failures arising from incorrect specification (including allocation of function); errors in the software; and electrical interference are all examples of systematic failures.

The safety integrity level for the safety-related systems is specified in terms of the three system elements - the exact package of which will depend upon the application in question and therefore the level of safety to be achieved. This package constitutes the safety integrity criteria for the application.

Design and Assessment Framework

The overall framework is shown in Fig 4.

It can be seen that:-

- The safety integrity criteria relate to the **TOTAL** configuration of safety-related systems. They therefore include Human Factor issues such as Allocation of Function.

- The safety integrity criteria are used as the basis of design and analysis of the safety-related systems.

It is intended that future guidance documents will specify the safety integrity criteria for specific applications. These criteria will include Human Factor issues, where appropriate.

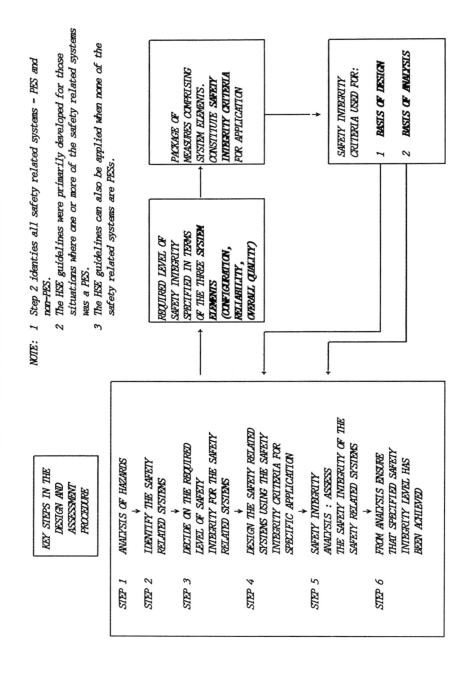

FIGURE 4 : OVERALL FRAMEWORK: DESIGN AND ASSESSMENT

Allocation of Function issues occur when considering the Configuration system element as the following example illustrates.

Consider the process shown in Fig 5. The vessel contains a liquid in which there would be serious safety implications if the level rose beyond a critical point. The basic system is as follows:-

(1) On the lower part of the vessel is a level transmitter feeding into the process computer. The signal from the level transmitter provides:-

- Control of the valve which itself controls the liquid in the vessel.

- Level indication for the process operator.

- An alarm set at a prescribed level, eg 80%. This alarm operates through the process computer software (ie it is software based).

(2) Above the level transmitter is a single level switch which is hardwired into an alarm.

(3) Above the single level switch are three level switches feeding into two programmable controllers (PC1 and PC2). Within each PC, 2 out of 3 voting takes place. The output of each PC goes to the trip valve and also to a trip alarm.

(4) A trip valve can also be operated by direct means from a stop button which is hard-wired.

From a control and protection viewpoint there are essentially four systems. (See Fig 6).

(1) As the level of the liquid rises in the vessel to say 80% level Alarm 1 is raised. The plant has been so designed (including incorporation of human factors principles) that the operator can on receipt of Alarm 1 take corrective action to

FIGURE 5

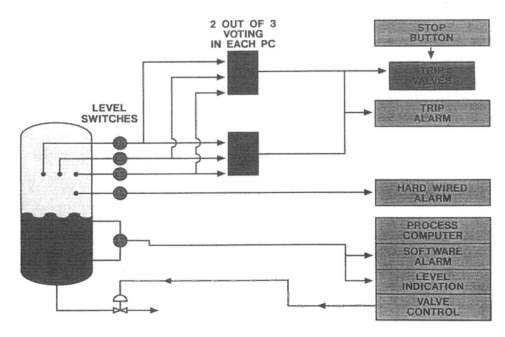

FIGURE 6

CONFIGURATION OF CONTROL AND PROTECTION SYSTEMS

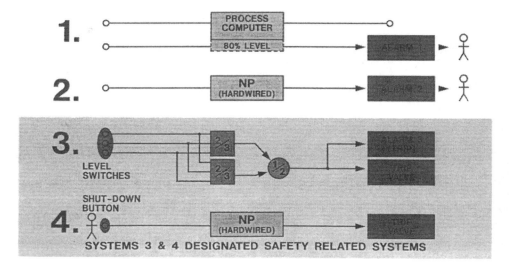

prevent the liquid level from rising any further. This is system 1.

(2)   If, however, the operator cannot control the liquid level or, the process computer is incapable of taking the corrective action even though the operator has performed correctly, the liquid level may rise until the hardwired Alarm 2 is raised. Even at this point the operator may be able to take corrective action to stop the liquid level rising any further. This is system 2.

(3)   If the operator cannot control the liquid level and the liquid continues rising, it will then activate, via the three level switches and PC1 and PC2, the trip valve to automatically bring the plant to a safe state. This is system 3.

(4)   Should system 3 fail, then the operator can bring the plant to a safe state by operating a hardwired system activated by a shut-down button. This is system 4.

In this particular case, the safety-related systems are systems 3 and 4. These 2 systems provide the requisite level of safety, taking into account factors such as the level of hazard involved and the number of demands made upon systems 3 and 4 by failure of systems 1 and 2.

It can be seen that whether a system is SAFETY-RELATED is determined by the particular circumstances and many factors need to be considered. The role of the operator in systems 1 and 2 is very important, since efficient and correct operator action will minimise the DEMANDS on the two protection systems (systems 3 and 4). This has an important bearing on the level integrity required of systems 3 and 4 in order to achieve an acceptable HAZARD RATE.

Correct operator action will depend upon, amongst other things, the allocation of functions to the operator. However, in this very simple example this is unlikely to be a problem as the operator would be in no doubt as to the state of the plant. In addition the emergency action is very straight forward requiring very little skill.

SAFETY AND ALLOCATION OF FUNCTION

A recent incident where control-room operators at a chemical plant were overwhelmed with plant alarms has highlighted again the lack of understanding in industry about human capabilities in absorbing and interpreting large amounts of information.

If we were designing such a plant from scratch an allocation of function analysis of the type shown in Fig 7 might well be carried out. By allocation of function we mean producing a man-machine mix that takes advantage of the qualities that both men and machines bring to systems. Its antithesis is the 'left-over' approach where only those tasks 'deemed' uneconomic or inappropriate for machines are left to people.

This Decision Analysis approach [3] has the advantage of being systematic but as with all such techniques is open to the criticism of bias in the judgemental process.

When carrying out an allocation of function analysis it is important to keep in mind that the ultimate aim from a safety viewpoint is to produce an OVERALL system that is:-

(1)   Inherently safe

(2)   Robust to human error

So that an acceptable level of overall risk is obtained.

Inherent safety [4] is a safety engineering concept that entails examining the process for potential hazards and then in considering solutions to mitigate the hazard applying the methods shown below.

Increasing
Preference

(1)   Substitution (Avoid)
(2)   Intensification (Use less of)
(3)   Attenuate (Use them under conditions that make them less hazardous)
(4)   Contain
(5)   Control
(6)   Survive

Figure 7

Allocation of Function

Stage 1:- Consider all possible ways a function can be implemented.

Stage 2:- Write a narrative description of the design alternatives so as to allow a qualitative comparison.

Stage 3:- Establish criteria for the entire system (eg Availability, safety, cost etc)

Stage 4:- Rate the various criteria to obtain their relative importance or weight - eg safety 10 times more important than availability - normalise the weights.

Stage 5:- Rate the alternatives on the criteria (perhaps using tables of relative merit as a guide?) and create a score by using the weights. The alternative with the 'best' score is selected.

Adapted from [8]

Inherent safety concentrates on 1, 2, and 3 and has an underlying philosophy of trying to make plant and processes as simple as possible (perhaps by rationalising and simplifying the existing control and instrumentation scheme). This has obvious implications for human factors: The simpler the system the easier it is for the operator to understand it.

To design a system to be robust to human error requires that we first define human error. A scheme devised within HSE classifies human errors as:-

Lapses of attention

The individual intentions and objectives are correct and he selects the proper courses of action but a slip occurs in performing it.

Mistaken actions

Doing the wrong thing under the impression that it is right.

Mis-perception

This may be unintentional (tunnel vision or pre-conceived diagnosis). In other circumstances uncomfortable information may be subconsciously suppressed.

Picking the wrong target

This addresses the potential conflict between safety and other objectives.

Wilfulness

This is rarely the cause of incidents, but sometimes occurs.

The above classification scheme is discussed further in HSE's forthcoming booklet on Safety and the Human Factor.

If allocation of function is to contribute to system safety then in postulating various design alternatives and evaluating them then there is need to consider the implications for the various types of human error described aove; and just as importantly the positive strengths people bring to system safety.

For example mis-perception played a part in a serious incident that occurred in a chemical plant when start-up "alarms" on a process caused operators and managers to ignore "real" alarms. Picking the wrong target (ie production over safety) was a key factor in an incident where automatic safety features were defeated because of operational difficulties.

These examples highlight an immediate difficulty with allocation of function: What types of human error does correct allocation alleviate? Many I would suspect would feel mis-perception errors (cognitive factors) were covered; but can "wrong target" errors for instance be foreseen at the design stage? To prevent that type of error a formal review of human factors in the operating plant at an early stage would seem the more realistic solution.

Simple systems

The liquid level system described in Section 3 has well understood failure modes. Although that example is perhaps somewhat over simplistic it should not be forgotten that many industrial plants are relatively simple.

Provided the man-machine interface has been designed according to human factors principles eg information display, control layout, lighting, noise etc, then the main focus for alleviating human error will probably lie in organisational and personnel spheres of influence rather than problems associated with, say, the operators' mental model of the plant.

One issue within the above spheres of influence that has some current prominence in the ergonomics literature concerns the quality of the job left for the operator/supervisor.

HSE's forthcoming guidance booklet stresses the importance of the correct safety climate within a firm and this is hardly likely to be good if jobs are made "machine-like" devoid of human needs.

As a way of overcoming this problem Sell [5] advocates the principle of upgrading jobs eg allowing operators of CNC machines to change tapes and tools (traditionally done by technical/maintenance staff) so as to make a whole coherent job. Winterton [6] also argues for the enhancement of workers skills by using a "human-centred" approach to new technology.

These and other strategies for Human Factors design are discussed further in HSE's booklet. Questions particularly relevant to PES design and assessment are given in a Checklist in the PES2 Publication (see Fig 8).

Complex Systems

These systems are characterised by either or both of the following:-

(1)   Plant response to multiple failures can be difficult to interpret.
(2)   Safe state not immediately reachable

Such systems are usually highly automated and the operator's role is as system manager ie decision maker and although many safety features are automated the operator is often given the role of dealing with "unforeseen" emergencies.

The key issue is how to design the overall system so that the operator does not make mistakes or misperceive system status through an inadequate 'mental model' of the process; or that he is so overloaded that he is unable to perform adequately.

FIGURE 8

Checklist No 1A : Safety requirements specification

Item No	Item				Comments
1A.1	Do safety requirements originate from a systematic hazard study of the plant and, if not, on what basis was the safety requirements specification formulated?	Y	N	NA	
1A.2	Is there a formal procedure to check the safety requirements specification against the known hazards?	Y	N	NA	
1A.3	Is there a clear and concise description of each safety related function to be implemented by the PES, eg interlocks, alarms, trips, input data validity checks?	Y	N	NA	
1A.4	Are the safety related functions defined for every operating state of the plant eg: start-up, normal operation, shutdown, maintenance? (For example, certain trips or alarms may be inhibited or may operate at different levels during start-up)	Y	N	NA	
1A.5	Are the safety related functions defined for every operating mode of the PES, eg: auto, manual, maintenance? (For example, different interlocks may apply in different modes).	Y	N	NA	
1A.6	Are the necessary conditions for a safety transition between the operating modes in item 1A.5 above adequately defined and are unsafe transitions inhibited?	Y	N	NA	
1A.7	(i) Have safety performance requirements (speed accuracy, etc) been specified where necessary?	Y	N	NA	
	(ii) Have they been adequately researched?				
1A.8	Are the PES inputs relating to each safety function defined with regard to range, accuracy, noise limits, bandwidth, etc?	Y	N	NA	
1A.9	Are the PES outputs relating to each safety related function defined with regard to range, accuracy, update frequency, etc?	Y	N	NA	
1A.10	Have safe plant states been identifed so that a (i) Has the PES sampling rate and accuracy been defined?	Y	N	NA	
	(ii) Is it consistent with the defined inputs and outputs?				
1A.11	Have safe plant states been identified so that a safe state can be achieved in the event of defined PES failures?	Y	N	NA	

Work in this area has mainly been centred around nuclear power plant control rooms [7].

The techniques for tackling the problem include:-

(1)   Make the plant simpler
(2)   Design the information systems to match the operators mental model of the process.
(3)   Provide alarm analysis.
(4)   Provide decision support aids.
(5)   Provide training through the use of realistic simulators to enhance diagnostic skills; to develop the operator's 'mental model' and to increase familiarity with emergency procedures

Many of these techniques are still under development and have unfortunately not found much application outside Nuclear Power Plants; although alarm analysis is beginning to appear on some chemical plant in the UK.

The balance to be made between automation of safety-related systems in complex plants and the use of human versatility and flexibility to deal with the unexpected is still a matter of some debate.

CONCLUSIONS

Within the context of the HSE's PES guidelines allocation of function occurs when considering the configuration of safety-related systems. All safety-related systems need to be assessed as a package and this includes "Human Systems".

Allocation of function is an important factor in achieving system safety. In considering design alternatives attention needs to be paid to both inherent safety concepts and to defences against the various forms of human error that can occur through the entire life-cycle of a plant. (ie design, construction, operation, maintenance ... etc).

Economic pressures will undoubtedly lead to increased automation of safety-related systems. The consequences of such a policy need to be reviewed at the design stage and the consequences evaluated for the whole of the human factors control strategy shown in Fig 1.

Although much progress has been made in that respect a number of difficult areas remain such as design for cognitive factors in complex plants. Another difficulty which unfortunately is present in large areas of the human factors discipline is the lack of definite evidence of the extent to which the various solutions suggested by human factors actually improves performance.

Such information is particularly important when considering medium to low hazard situations where it would not be "reasonably practicable" to apply state of the art solutions.

As stated at the beginning of this paper The Control Systems Group of Technology Division in HSE is currently reviewing this particular topic and considering the way forward.

Possibilities include:-

Collaborative studies with industry
Development of application-specific standards
Published guidance
Targeted research

The Control Systems Group would very much welcome your views and comments.

CROWN COPYRIGHT

REFERENCES

1. Programmable Electronic Systems in Safety Related Applications: "1 An Introductory Guide" (ISBN 0118839136), HMSO PO Box 276, LONDON SW8 5DT.

2. Programmable Electronic Systems in Safety Related Applications: "2 General Technical Guidelines" (ISBN 0118839063), HMSO, PO Box 76, LONDON SW8 5DT.

3. Edwards W, Decision Making. In handbook of Human Factors ED, G Salvendy. Wiley 1987.

4. Kletz T A. Cheaper, Safety Plants or Wealth and Safety at Work. Institution of Chemical Engineers, Rugby Warwickshire, CV21 4HQ, England.

5. Sell R G. The Changing Role of the Operator. In Occupational Health and Safety in Automation and Robotics. Ed K Niro. The Proceedings of the 5th U O E H International Symposium Kital Kyuski Japan, 20-21 September 1985, Taylor and Francis 1987.

6. Winkerton J, New Technology and Redesigning Work. In Contemporary Ergonomics 1987. Ed E D Megaw. Proceedings of the Ergonomics Society's 1987 Annual Conference, Swansea Wales, 6-10 April 1987. Taylor and Francis 1987.

7. Woods D D. Human Factors Challengers in Process Control: The Case Study of Nuclear Power Plants. in Handbook of Human Factors. Ed G Salvendy. Wiley 1987.

8. Kantowitz B Allocation of Human Functions. In Handbook of Human Factors Ed G Salvendy. Wiley 1987.

AES/TD-004/B-4/6.88/SHH

Human Reliability Assessors Guide

An Overview

P Humphreys

Safety & Reliability Directorate

Synopsis.

The Human Reliability Assessors Guide l provides a review of
eight major techniques currently available for the quantification
of Human Error Probabilities. This Guide is a corporate venture
by the Human Reliability Assessment Group (HRAG), which is
sponsored by the United Kingdom Atomic Energy Authority.
The group comprises members from a wide variety of industrial,
regulatory and consultancy organisations, all having an active
interest in the furtherance of human reliability assessment.

Although some documentary evidence is available describing the
main techniques available, the style of presentation is often in
a format which is unsuitable for those wishing to apply the
technique. As a consequence the analyst has no clear picture of
the merits and limitations of the techniques, nor of the range of
problems to which the technique may be satisfactorily applied.

The Guide has two main objectives. The first is to provide a
clear and comprehensive description of eight major techniques
which can be used to assess human reliability. This is
supplemented by case studies taken from practical applications of
each technique to industrial problems. The second objective is
to provide practical guidelines for the selection of techniques.
The selection process is aided by reference to a set of criteria
against which each of the eight techniques have been evaluated.

Utilising the criteria and critiques, a selection method is
presented. This is designed to assist the potential user in
choosing the technique, or combination of techniques, most suited
to answering the users requirements.

For each of the eight selected techniques, a summary of the
origins of the technique is provided, together with a method
description, detailed case studies, abstracted case studies and
supporting references.

The guide provides the analyst with a realistic evaluation of the eight techniques, enabling the choice of techniques for any particular application to be made in a more logical and cost effective way.

1. Introduction.

Can we quantify human error? What quantification techniques are available to the analyst and are the available techniques adequate for the purpose?

In the light of major incidents such as Chernobyl, the Challenger space shuttle disaster, Bhopal, and the Kings Cross Underground station fire, there is evidently a need to provide a better understanding of the interactions between man, in his role as designer and operator of complex equipment, and the systems under his control. Certainly, the quantification of the probability of those human errors which could have catastrophic consequences is an objective which those involved in human error analysis have strived to achieve. But how successful are the approaches which are currently available, and how much effort and expense is needed to apply the techniques?

The Human Reliability Assessors Guide was conceived in order to provide some of the answers, and to act as a guide to the use of eight of the techniques currently in use today.

The Guide has two main objectives.

> The first objective is to provide guidelines for the selection of techniques. This is the content of Part 1. This first objective is achieved by defining a comprehensive set of criteria and then evaluating the techniques on these criteria. Based on these evaluations, a method is provided to assist the potential user in choosing a technique or combination of techniques for the problem of interest.

> The second objective is to present a comprehensive description of eight of the major techniques used in the assessment of human reliability in the context of probabilistic safety and risk assessment, together with case studies illustrating their application to practical problems. This is the content of Part 2 of the guide.

The techniques selected were:-

    A. Absolute Probability Judgement (APJ) [2]

    B. Paired Comparison (PC) [3]

    C. Technica Empirica Stima Errori Operatori (TESEO) [4]

    D. Technique for Human Error Rate Prediction (THERP) [5]

    E. Human Error Assessment and Reduction Technique (HEART) [6]

    F. Influence Diagram Approach (IDA) [7]

    G. Success Likelihood Index Method (SLIM) [8]

    H. Human Cognitive Reliability Method (HCR) [9]

The techniques were selected for inclusion on the grounds that they had been applied to real industrial problems, and that case study material was readily available to the authors exemplifying the use of the techniques. A second criterion was that the techniques were well documented such that their underlying theoretical rationales and methods of application could be adequately described.

1.2 Industrial applications.

The industrial applications which the subject matter covers are indicated in table 1.

2. The reporting of the techniques.

The Guide presents a brief summary of each technique, as illustrated in Table 2 for THERP. This is followed by an extended method description and then a brief statement of the advantages and disadvantages of the technique. Some examples of the disadvantages and advantages of THERP as reported in the Guide are shown in tables 3 and 4 respectively. Case studies and extended abstracts are included wherever possible to complete the picture for each technique. To assist those users who wish to obtain further information on a particular technique, extensive references are provided.

Industrial Area

	Nuclear	Chemical	Offshore	Service	Transport	Defence	Mining	Experimental Research
Risk Assessments	D2, D3 D5, F2 H2	C2, E2 G3	A3, A5 F3		B2	E4		
Design Studies	D2	E2	A3, A5	A2				
Evaluation Studies	B3, D4 G2	C2, E3		A2	B2			A4, B4
Emergency Operation	D2, D5 F2, G2	C2	A5, F3		B2			
Process Control/ Normal Operations	D2	C2, E2 G3				E4	G4	
Maintenance	D3							

Table 1   Part 2 Case Study Content
(The letter in each cell refers to a technique and
the number to a case study in Part 2 of the Guide.)

D. TECHNIQUE FOR HUMAN ERROR RATE PREDICTION (THERP)

The Technique for Human Error Rate Prediction (THERP) was developed by Swain and Guttmann of Sandia Laboratories for the US Nuclear Regulatory Commission (Swain and Guttmann[5]) and is the most widely used technique to date. It is basically a hybrid approach, since it models human errors using probability trees and models of dependence (i.e. as in engineering risk assessment), but also considers Performance Shaping Factors (PSF) affecting the operators actions. The technique dates back to the concept of a data store of human reliability (Altman[10]) and in fact the technique is dependent on a data base of an error probabilities contained in chapter 20 of the THERP handbook. These performance data were derived from a mixture of data and judgement by the authors of the technique. This data base, whether valid or not, makes it unique amongst HRA techniques. This, together with its engineering approach, and the fact that it was the first major methodology to be accepted and used in the field, almost certainly accounts for the popularity and frequent usage of THERP. THERP has also been the subject of a great deal of scrutiny, particularly in the peer review study abstracted in D4.

The main case study refers to a British assessment carried out in the process control industry, using a variant of THERP, which is sufficiently close to the original method to make it a useful demonstration of the power and limitations of the THERP methodology. The second case study refers to an analysis of a maintenance testing procedure which uses the THERP approach to modelling errors and their potential recovery, although the THERP data base itself is not employed. Thirdly, there is an abstract of the peer review of THERP, which discusses some of the problems and disadvantages with this technique. Finally, an abstract is presented of a Probabilistic Risk Assessment which evaluated a series of actions in a nuclear power plant emergency scenario.

Table 2. Example of technique summary.

Disadvantages of THERP

Resource requirements considerable: A very detailed THERP analysis can require a large amount of work to generate HEP values. In practice by tailoring the detail for the THERP analysis to the appropriate level for each job, the amount of work can be minimised.

Diagnosis and High Level Decisions: THERP is not particularly good for evaluating errors concerned with diagnosis or high level decisions (i.e. those requiring significant conscious thought).

Does not produce explicit design recommendations: THERP is a relatively crude human reliability evaluation tool, best used in support of engineering assessments. It is not an 'ergonomic tool' in that it is not particularly sensitive to many possible performance shaping factors nor does it attempt to identify the underlying psychological causes of a particular error. Expert advice should be used to determine the more subtle 'ergonomic' factors that can improve human reliability.

THERP peer review: Several problems were encountered during the peer review study abstracted in D4, particularly relating to a lack of uniformity in the 'modelling' of the scenarios and in the use of the data base. This led in some cases to very large discrepancies in different analysts estimates for the same task.

Table 3. THERP disadvantages.

Advantages of THERP

Comparisons - Design and Risk Assessments: THERP can be used easily for design, risk and reliability assessments at all stages of design. The level of detail of the THERP analysis can be tailored according to the depth of the overall assessment being performed e.g. early design conceptual assessments would only require a crude THERP analysis. Thus, the resulting HEPs are of a similar quality to the other data used in the comparisons.

Integration into PRAs: THERP can be integrated into Probabilistic Risk Assessments easily. Its form and approach make it compatible with fault tree methods.

Scrutable: THERP provides a structural, logical, well documented record of the factors and errors considered in the human reliability assessment. This allows the results to be reviewed easily and assumptions used to be examined. This can be particularly important in sensitive areas (for safety or design changes).

THERP is generally acceptable on several different human reliability domains and is high on 'face validity'.

Use of alternative sources of data: THERP can be used either with the handbook data or any other data that is considered to be relevant, e.g. simulator data; known plant-specific data; or data produced from other human assessment methods for a particular task.

Table 4. THERP advantages.

3. Criteria for evaluation of the techniques.

Each of the eight techniques was evaluated against six criteria. The criteria were selected both as an aid to technique evaluation in a comparative way and, as a means of aiding the selection of one or more techniques to perform an analysis.

The criteria against which each technique was evaluated are :-

Accuracy
   numerical,
   consistency.

Validity
   modelling,
   theoretical,
   perceived,
   comparative.

Usefulness
   qualitative,
   sensitivity analysis capability,
   breadth of applicability,
   comprehensiveness.

Effective use of resources
   equipment and personnel,
   data requirements,
   resource limitations,
   degree of decomposition,
   training requirements.

Acceptability
   to regulatory bodies,
   to the scientific community,
   to assessors,
   auditability,
   expert review.

Maturity
   current,
   development potential.

The selection and definition of the six criteria was certainly a headache for the review team, and certainly the reader is at liberty to disagree with the definitions finally chosen, but the criteria provided the review team with a means of evaluating the eight techniques with a high degree of consistency. Certain of the criteria such as Numerical Accuracy, were a particular problem, since, while in the perception of many users the predictive numerical accuracy of a technique will be the major consideration, the accuracy criterion is the most problematic of all the criteria to apply.

The difficulty of validating the numerical accuracy of techniques stems from the requirements for empirical data with which to compare their predictions. The three major potential sources of reference data are

> operational experience,
> simulators, and
> experimental literature on human performance.

Operational experience, has not yielded very much freely usable data. Simulators have considerable potential to generate reference data. Unfortunately, the running costs of these simulators are extremely high, and they are used almost exclusively for training purposes. The experimental literature on human performance needs to be analysed if it is to be used as a source of validation data. Thus, until such data are available, the determination of accuracy and validity remains a subjective issue and open to debate and other criteria may be better used to select techniques.

To aid the reader of the guide in understanding the meaning behind each criteria, in addition to providing a formal definition of each criteria, the authors have included a commentary with each of the subcriteria and two examples are given here.

Validity

Modelling validity. The degree to which the technique explores, elicits, and incorporates modelling and general information regarding factors influencing human reliability.

This subcriterion reflects the extent to which the technique can take into account any contextual information which may be available e.g. the task characteristics, the context within which it is carried out or the attributes of the team or the operator who performs the task. It is also important as it addresses the qualitative aspects of human error assessment.

Usefulness

Comprehensiveness. The range of task types, behaviours and types of mental processes that the technique can be applied to.

In traditional forms of risk assessment, human reliability techniques have been applied primarily to simple proceduralised tasks. However, there is an increasing tendency to also consider diagnostic and rule following errors in Probabilistic Safety Assessments (PSA), and therefore the capability of a technique to handle all types of human performance encountered in PSA will lead to a high rating on this criterion.

4. Method evaluation.

Each of the eight techniques was evaluated and the conclusions presented under the respective criteria, so that the reader of the Guide might compare each technique on a single criterion. A subjective five point scale was used to rank the success of each technique against each criterion as:-

> low,
> low to moderate,
> moderate,
> moderate to high,
> high.

A summary of the evaluation is given in table 5.

	APJ	PC	TESEO	THERP	HEART	IDA	SLIM	HCR
Accuracy	Moderate	Moderate	(Low)	Moderate	Moderate	(Low)	Moderate	(Low)
Validity	Moderate/High	Moderate	Low	Moderate	Moderate	Moderate	Moderate	Low
Usefulness	Moderate/High	Low/Moderate	Moderate/High	Moderate	High	Moderate/High	High	Low/Moderate
Effective Use of Resources *	Moderate	Low/Moderate	High	Low/Moderate	High	Low/Moderate	Low/Moderate	Moderate
Acceptability	Moderate	Moderate/High	Low	High	(Moderate)	(Moderate)	Moderate/High	Low/Moderate
Maturity	High	Moderate	Low	High	Low/Moderate	Low/Moderate	Moderate/High	Low

* A rating of High on this criterion means the technique is favourable with respect to effective use of resources (i.e. resource requirements are low), and vice versa.

(ranking) these rankings are based on the subjective opinions of the authors and contributors to the guide, since insufficient evidence was available to justify a ranking from applications alone.

Table 5. Summary of Evaluations.

5. Guidelines for the Selection of Techniques.

One of the objectives of the Guide was to provide selection criteria to enable a potential user of the techniques to select one or more techniques to suit their needs. Three approaches are presented in the guide.

The first approach, Informal Evaluation, simply requires the potential user to identify which criteria are the most important for their application, and then to select the technique(s) which score highest on those criteria. This simplistic approach should be used only as a last resort.

The second approach, Decision Analysis, is based on the Systematic Multiattribute Rating Technique. The stages of the process are :-

> rank the criteria in descending order of importance in relation to the application,

> assign the least important criterion an arbitrary value of 10,

> use judgement to decide how much more important the next criterion is,

> repeat the assignment until all criteria have been allocated a numerical score relative to the least important criterion,

> normalise the criteria by dividing the value for each criterion by the sum of all the values.

Note that it is essential that the process is repeated for each new application, since priorities on subsequent projects may be different.

An example of the process as applied to a hypothetical project is shown in tables 6, 7 and 8, for the Absolute Probability Judgement (APJ) technique. The process should be repeated for each technique.

The third approach, Selection Matrix, directs the potential user of the techniques to ask certain questions which will assist in eliminating those techniques which are unsuitable for the application in hand. Fourteen such questions are provided and examples of some of the questions and responses are presented in table 9. It should be recognised by the potential user of the selection matrix that:-

> a)   not all questions and responses may be important for every application and,

> b)   the list is not exhaustive.

Criteria (in descending order of importance)	Relative Weights (RW_i)	Normalised Weights (NW) $(RW_i / \quad RW_i)$
Accuracy	40	0.30
Validity	30	0.22
Usefulness	20	0.15
Effective use of resources	20	0.15
Acceptability	15	0.11
Maturity	10	0.07
	—	——
	$RW_i = 135$	$NW_i = 1.00$

Table 6. Example of derivation of normalised weights for criteria.

	APJ	PC	TESEO	THERP	HEART	IDA	SLIM	HCR
Accuracy	3	3	1	3	3	1	3	1
Validity	4	3	1	3	3	3	3	1
Usefulness	4	2	4	3	5	4	5	2
Resources	3	2	5	2	5	2	2	3
Acceptability	3	4	1	5	3	3	4	2
Maturity	5	3	1	5	2	2	4	1

Table 7.   Numerical Ratings of Techniques.

Note, the ratings of table 7 are derived in accordance with the following :-

1 = low, 2 = low/moderate, 3 = moderate,   4 = moderate/high and 5 = high.

Criterion	APJ rating x normalised weight			
Accuracy	3	0.30	=	0.90
Validity	4	0.22	=	0.88
Usefulness	4	0.15	=	0.60
Resources	3	0.15	=	0.45
Acceptability	3	0.11	=	0.33
Maturity	5	0.07	=	0.35
				——
		Preference Index	=	3.51

Table 8. Example Calculation of Preference Index for APJ.

	APJ	PC	TESEO	THERP	HEART	IDA	SLIM	HCR

Y = YES, N = NO

Is the technique applicable for:

	APJ	PC	TESEO	THERP	HEART	IDA	SLIM	HCR
Simple and Proceduralised Tasks?	Y	Y	Y	Y	Y	Y	Y	N
Knowledge-Based, Abnormal Tasks?	Y	Y	Y	N	Y	Y	Y	Y
Misdiagnosis which makes a situation worse?	Y	Y	N	N	N	Y	Y	N
Are Qualitative Recommendations Possible?	Y	N	Y	Y	Y	Y	Y	Y
Is Sensitivity Analysis Possible?	N	N	Y	Y	Y	Y	Y	Y

Does the technique have:

	APJ	PC	TESEO	THERP	HEART	IDA	SLIM	HCR
Requirements for Calibration data?	N	Y	N	N	N	N	Y	N
Requirements for experts (judges)?	Y	Y	N	N	N	Y	Y	N

Table 9. Selection Matrix.

6. Conclusions.

The Human Reliability Assessors Guide provides a peer review of eight of the main techniques in use today for the quantification of human error.

By reading the method descriptions and annotated examples describing the application of each technique, the reader can become familiar with the techniques and the benefits and pitfalls associated with the techniques.

Through the application of the method selection approaches, the potential user can remove some of the uncertainty normally attached to technique selection.

Finally, in the production of the guide it has become evident that there is seldom a time when just one technique will satisfy all of the requirements of a project. It is advisable to utilise the capabilities of two or more techniques to more specifically address the problem of human error analysis.

7. Acknowledgements.

The author wishes to thank those people and companies who have been instrumental in creating the guide and supporting its development. Particular thanks go to the principal authors David Embrey, Barry Kirwan and Keith Rea and to all members of the HFRG sub group who have provided peer review and contributions to the guide.

8. References.

1. Human Reliability Assessors Guide. Human factors in Reliability Group. To be Published by NCSR, UKAEA, Wigshaw Lane, Culcheth, Warrington. 1988.

2. COMER, M.K., SEAVER, D.A., STILLWELL, W.G., and GADDY, C.D. (1984). Generating Human Reliability Estimates Using Expert Judgement, Vol I. Main Report. NUREG CR-3688/10F2. 5 and 84-7115, GP-R-213022.

3. BLANCHARD, P.C., MITCHELL, M.B., and SMITH, R.L. (1966) Likelihood-of-accomplishment scale for a sample of man-machine activities. Dunlap and Associates, Inc., Santa Monica, CA, USA.

4. BELLO, G.C. and COLOMBARI, V. (1980). The Human Factors in Risk Analyses of Process Plants: The Control Room Operator Model, "TESEO". Reliability Engineering, 1, pp 3-14.

5. SWAIN, A.D. and GUTTMAN, H.E. (1983). Handbook of Human Reliability Analysis with Emphasis on Nuclear Plant Applications. NUREG/CR-1278, U.S. Nuclear Regulatory Commission, Washington, D.C.

6. WILLIAMS, J.C. (1986) HEART - A Proposed Method for Assessing and Reducing Human Error, in Proceedings of the 9th Advances in Reliability Technology Symposium, University of Bradford, 4 April.

7. HOWARD, R. and MATHESON, J.G. (1980), Influence Diagrams SRI International, Menlo Park, California, U.S.A.

8. EMBREY, D.E., HUMPHREYS, P.C., ROSA, E.A., KIRWAN, B., and REA, K. (1984). "SLIM-MAUD: An Approach to Assessing Human Error Probabilities Using Structured Expert Judgement". NUREG/CR-3518, (BNL-NUREG-51716). Department of Nuclear Energy, Brookhaven National Laboratory, Upton, New York 11973, for Office of Nuclear Regulatory Research, US Nuclear Regulatory Commission, Washington, D.C. 20555.

9. HANNAMAN, G.W., SPURGIN, A.J. and LUKIC, Y.D. (1984) Human Cognitive Reliability Model for PRA Analysis, Draft Report NUS-4531, EPRI Project RP2170-3, Electric Power and Research Institute, Palo Alto, California.

10. ALTMAN, J. W. (1967).Classification of Human Error. In W. B. Askren (Ed) Symposium on reliability of human performance in work (Report No. AMRL-TR-67-88). Wright-Patterson AFB, OH: Aerospace Medical Research Laboratories, 5-16.

A Comparative Evaluation of Five Human Reliability Assessment Techniques

B Kirwan
BNFL, Safety Department

ABSTRACT

A field experiment was undertaken to evaluate the accuracy, usefulness, and resources requirements of five human reliability quantification techniques (THERP; Paired Comparisons, HEART, SLIM-MAUD, and Absolute Probability Judgement). This was achieved by assessing technique predictions against a set of known human error probabilities, and by comparing their predictions on a set of five realistic PRA human error scenarios.

On a combined measure of accuracy THERP and Absolute Probability Judgement performed best, whilst HEART showed indications of accuracy and was lower in resources usage than other techniques. HEART and THERP both appear to benefit from using trained assessors in order to obtain the best results. SLIM and Paired Comparisons require further research on achieving a robust calibration relationship between their scale values and absolute probabilities.

Author's Acknowledgement: This work took place as a result of the influence of Mr G Hensley, recently retired from BNFL (who it was originally intended would co-author the paper), who has been involved in this difficult area for the past eleven years, working towards the development of sound and usable methods.

Disclaimer: This paper, its results and conclusions, are solely the opinions of the author and do not necessarily represent those of BNFL.

INTRODUCTION

The effects of human error in accident sequences of such catastrophic events as Bhopal, Chernobyl, and Three Mile Island, has focussed greater attention on techniques to predict human error for inclusion in probabilistic risk assessments (PRA). At present, however, opinions differ on the most appropriate model for this task. There are several techniques available, but few definitive guidelines on which technique or set of techniques is best. This is because validation studies have largely been inconclusive, and only a few comparative evaluations have been performed. This poses a problem for those industries that make use of PRA's to demonstrate adequate plant safety, eg the nuclear, chemical and offshore industries, in that whilst accurate quantification of human reliability is desirable, there are many limitations in existing human error assessment models.

A recent review of eight such models[1] suggested that five of the models in particular might be accurate and useful in a PRA context. These techniques were:

THERP (Techniques for Human Error Rate Prediction)
HEART (Human Error Assessment and Reduction Technique)
SLIM (Success Likelihood Index Method)
APJ   (Absolute Probability Judgement)
PC   (Paired Comparisons)

The major limitation with the qualitative review of techniques was that there was insufficient evidence on the most obvious criterion for the selection of techniques, namely accuracy, to make a conclusive statement about which techniques were best for use in PRA. This led the review to assess the techniques on a set of criteria, including accuracy, based on the limited information which was available, and evaluating the validity and qualitative usefulness of each technique (eg, the degree to which it aided error reduction following quantification). The resource requirements of each technique, the general acceptability of the technique (eg, to the PRA and scientific community) and the maturity of the technique, were also assessed in this review.

This qualitative review generated a great deal of useful information and discussion upon which selection of techniques could be based. Within BNFL Engineering Division, it was decided to carry out a comparative assessment of the five 'best' techniques from that review, to investigate further the accuracy of the techniques, examining at first hand the performance of the techniques on a subset of the criteria mentioned above. The overall goal was to consider ways of improving BNFL's own approach to the quantification of human error probabilities in safety cases. BNFL's own approach (called simply Human Error Database, HED) was also investigated as part of the experiment, although the main goal of the experiment was the analysis of published and well-documented techniques. The results, nevertheless, also describe this method's performance. The evaluation of these techniques was achieved by carrying out a field experiment, and the rationale underlying this experiment and its limitations and hypotheses, are defined below.

Criteria for the Evaluation of Techniques

Six criteria were utilised in the peer review[1] to assess techniques, namely

Accuracy
Validity
Usefulness
Resources Usage
Acceptability
Maturity

Three of these, validity, acceptability and maturity are specifically not addressed in the experiment as they are assessed adequately in the peer review. The three remaining criteria which were addressed by the experiment are discussed below.

Accuracy

The accuracy of a technique, defined as the degree to which the predictions of a technique agree with the observed occurrence of errors, is the most obvious criterion upon which to assess techniques of human error quantification. However, as has been well documented, [1,2] there are few data upon which to make such comparisons, and this in part has led to a lack of necessary validation experiments. In this study, it was decided to gather together what data existed (the data set comprised 21 Human error probabilities (HEPs) spanning five orders of magnitude), and use them to 'validate' the techniques. Much of the data, unfortunately, are of a poorer quality than is desirable for such an important validation issue, originating from experimental studies on human performance. It is questionable therefore, whether such a data set is homogeneous with the type of assessments actually carried out in PRA. Nevertheless, currently these data are all that are available and, even though the data set is a mixture of experimental, simulator, industrial and synthesised data, it has been used to validate the identified techniques until better data are collected/derived.

The first hypothesis therefore, was that one or more techniques would accurately predict these data. A sufficient measure of accuracy (in terms of precision) is that the estimates from a technique are within a factor of ten of the actual data-based human error probabilities. In a validation experiment this represents a reasonable degree of precision and, given the uncertainties within risk assessment, this degree of precision in a technique will normally be adequate for PRA-usage.

Another component of the accuracy criterion is consistency, since for a technique to be useful and acceptable in PRA, it must be able to give results which are consistently accurate. **The second hypothesis, was therefore that two or more independent trials of the technique against the same data would yield approximately the same results.**

An alternative measure of accuracy, and arguably less robust a measure than the first definition of accuracy, is convergent or comparative validity. Convergent validity is the degree to which one technique's predictions on a particular scenario agree with another's. With good quality data available to validate techniques there is no need (beyond academic interest) to measure convergent validity. However, because some of the data were experimentally-based, and because they were not put within a PRA context (ie, within event and fault trees), it was decided to measure convergent validity. This was achieved by defining five realistic human error scenarios, typical of those used in safety cases (a mixture of fault and event trees), in a very realistic PRA-type context, and by determining convergent validity. **The third hypothesis was therefore that one or more pairs of techniques predictions would agree on one or more of the scenarios.**

The fourth hypothesis was that techniques validated by the database would also agree with each other significantly in the scenarios. If this was not the case, then it would suggest that the data set was not necessarily homogeneous with PRA-type assessments.

Another part of the accuracy criterion is the existence of a robust relationship between a technique's estimates and observed data. Thus, if a technique's estimates are not precise (within one order of magnitude of the actual values), but nevertheless are consistently related to the results, eg all are two orders of magnitude lower than the real estimates, then clearly although the technique is inaccurate, it has some predictive utility, and is perhaps merely 'miscalibrated'. This may in fact be seen as a measure of the validity of the technique. It is of particular interest with respect to those techniques which require calibration (PC and SLIM), since by definition those techniques could have predictive validity and utility but simply be miscalibrated. A fifth hypothesis therefore, was that one or more techniques predictions would be significantly related to the real values, irrespective of whether these predictions were precise, ie that there would be a statistically significant relationship between a technique's predictions and observed data.

Usefulness

One aspect of usefulness is comprehensiveness, ie, the degree to which a technique can accurately quantify all types of human error identified for quantification in a PRA (usually embedded within fault and/or event trees). Thus, for example, a technique should be able to quantify 'routine' slips and lapses (eg, closing the wrong valve or forgetting to close it), as well as more complex errors such as misdiagnosis. To evaluate fully the comprehensiveness of techniques requires a relatively large data set with many different types of errors. Within this experiment it was not possible to fully evaluate techniques on this criterion, due to a lack of such data, and overall resources limitations on this experiment. The experiment tried instead to account for this criterion by using different types of errors in the data set (ie, some routine and some problem-solving tasks), and by using five scenarios which contain most types of error encountered in current PRAs (including misdiagnosis and even the emergency use of a non-proceduralised but viable recovery strategy, ie a knowledge-based solution). Thus, although techniques are not specifically evaluated on the comprehensiveness criterion, in order to gain overall accuracy and convergence a technique must be relatively comprehensive, since a failure to be so will prevent it being accurate on all tasks in the experiment and degrade its overall performance on the accuracy criterion. Hence comprehensiveness, or rather lack of it, is measured indirectly in the experiment.

The major usefulness criterion evaluated in this experiment, given that comprehensiveness could not be assessed, was qualitative usefulness, ie, the degree to which a technique provides useful information on how to cost-effectively reduce human error probabilities if required (ie, by changing the operator's task, quality of the operator interface, procedures etc). The qualitative review[1] showed that SLIM and HEART in particular were highly useful techniques in this sense. Less clear was the usefulness of the other three techniques assessed in this experiment. The sixth hypothesis was therefore that one or more of the remaining three techniques (APJ, PC, THERP) would provide qualitatively useful information.

Resources

A major criterion, given approximately equal accuracy ratings for two or more techniques, is the degree of resources required to utilise a technique. Whilst some techniques (notably SLIM) require a computer program, the major operational definition of resources is the number of personnel days required to use the technique, since for virtually any high risk technology organisation this will dominate the cost of application of the technique. Also, if a technique is to be utilised for a long time in a company, initial and relatively small computer costs are very quickly outweighed by personnel costs.

Personnel costs break down into three parameters, namely the assessor's time, the length and depth of training in the use of the technique (where required), and the number of 'expert' days required, which includes the difficulty of convening expert panels (ie, finding experts and co-locating them for an expert panel session) if required. Resources is thus a composite measure. The seventh hypothesis was therefore that some techniques would cost less in resources than others. Note that this criterion must be considered in conjunction with other criteria (eg, usefulness) since, for example, a technique may be high on personnel resources but produce a great deal of useful information, in which case it may be a resource-effective technique. However, since organisations will differ as to whether information beyond mere numbers is useful, the experiment simply evaluates resource usage, and not resources effectiveness (which must be considered by the individual organisation).

As training time is obviously of particular interest to any reasonably large safety assessment group, it was decided to look at training requirements for the non-expert-group techniques ie, THERP and HEART. This was achieved by using two proficient and one novice HEART/THERP assessor, and seeing whether the novice was as good as the assessors with greater experience of the techniques. Whilst this use of a single novice cannot definitively answer the question of whether it is necessary to use trained assessors, it was felt that at least an indication could be gained as to whether training appeared to help. Thus, the eighth hypothesis was that novice users would not be as accurate as proficient HEART/THERP assessors.

Techniques Assessed in the Experiment

The five techniques which were chosen for the evaluation are described briefly below (for a fuller description see reference 1, or the source references given with each technique).

THERP: Technique for Human Error Rate Prediction [3]

This technique basically comprises a database of human error probabilities, performance shaping factors such as stress which affect human performance and can be used to alter the basic human error probabilities in the database, an event tree modelling approach, and a dependency model. It is often referred to as being a 'decompositional' approach, in that its description of human operator tasks has a higher resolution than many other techniques,[4] but nevertheless it is a logical approach with a larger degree of emphasis on error recovery than most other techniques.

SLIM-MAUD: Success Likelihood Index Method Using Multi Attribute Utility Decomposition

SLIM-MAUD, is a computerised technique originating from the world of decision analysis. SLIM is essentially a method of defining preferences amongst a set of items, in this case human error tasks. MAUD is a sophisticated approach which helps ensure that the expert group, used to define the preferences, do so without letting biases affect their judgements[6]. SLIM-MAUD defines preferences which represent the relative likelihoods of the errors , as a function of various factors which can affect human performance (Performance Shaping Factors (PSF), eg, level of training, quality of the procedures, time available, quality of the operator interface, etc). SLIM-MAUD creates a relative scale representing the likelihoods of the errors, called the Success Likelihood Index (SLI). This index can be "calibrated" to generate human error probabilities via a logarithmic relationship based on experimental data[7] and a theoretical argument[8] presented on behalf of the Paired Comparisons Technique (see below).

APJ: Absolute Probability Judgement [9]

APJ is the use of experts to generate human error probabilities directly. It may occur in various forms, from the single expert assessor, to the use of a large group of individuals who may work together, or whose estimates may be mathematically aggregated. APJ requires experts, and these experts must firstly have substantive expertise, ie, they must know in depth the area they are being asked to assess. A second requirement is that the experts must have normative expertise, ie they must be familiar with probability calculus, as otherwise they will not be able to express their expertise coherently in a quantitative form. It is preferable, if experts are meeting and sharing their expertise and discussing their arguments in a group, to utilise a facilitator. The facilitator's role is to try to prevent biases due to personality variables in the group, and biases in making expert judgements, from occurring and distorting the results.

Paired Comparisons [8]

Paired Comparisons is a technique borrowed from the domain of psychophysics (a branch of psychology), and similarly to SLIM, is a means of defining preferences between items (human errors). Paired Comparisons asks the experts to make fairly simple judgements. Each expert individually compares a pair of error descriptions and decides which error is more probable. For 'n' tasks, each expert makes $n(n-1)/2$ comparisons. When comparisons from different experts are combined, a

relative scaling of error likelihood can be constructed. This is then
calibrated using a logarithmic calibration equation, which, as for SLIM,
requires that the HEPs are known for at least two of the errors within
the task set. Paired comparisons is relatively easy for the experts to
carry out, and the method usefully determines whether each expert has
been consistent in the judgements made (inconsistency suggesting a lack
of substantive expertise).

HEART: Human Error and Assessment and Reduction Technique [10]

HEART is a relatively quick technique to use, and is based on a
review by its author of the human factors literature and, in particular,
of the experimental evidence of various parameters' effects on human
performance. The technique defines a set of generic error probabilities
for different types of task, and these are the starting point for a HEART
quantification. Once a task is thus classified it is determined by the
analyst whether any HEART error-producing conditions (EPC) are evident in
the scenario under consideration. For each EPC evident, the generic HEP
is multiplied by the EPC multiplier, increasing the HEP. The technique
also has a set of practical error reduction strategies which can be used
to reduce the impact of the error on the system, or prevent it entirely.

HED: Human Error Database

This method, called simply 'Human Error Database' (HED) was basically
derived from WASH 1400 [11] and safety assessors' judgement, and has been
used in some BNFL safety case human error assessments. Safety Department
currently use a predominantly fault-tree-based approach, and any operator
errors in the fault tree must be referenced either to a similar item in
the data base, or else this new HEP and its description must be entered
within the data base. Thus the data base contains simply a set of error
definitions and associated HEP values. The method has similarity with
THERP, since it is a data base derived from WASH-1400 type values, but it
is by no means as decompositional as THERP and has no specific model of
dependency (although human 'common mode' failure limits are often used in
safety cases to account for dependency in fault trees). The experiment
therefore analysed six methods, including the HED approach. It was of
natural interest also to determine how the HED method compared with the
others, and in particular to note whether it was optimistic or
pessimistic by comparison.

Modelling and Quantification

It has already been noted above that different techniques use
different levels of behavioural description in scenario analysis, THERP
using a high resolution, and techniques such as SLIM-MAUD often
quantifying a whole scenario (ie, a whole sequence of human actions) as
one HEP. When comparing methods of quantification, one possible source
of experimental confounding is that if two techniques disagree on the
same scenario, it will not be clear if they disagree because of the way
they have independently modelled the scenario, or because of fundamental
differences in quantification approaches. In this experiment therefore,

it was decided to prevent this possible confounding by pre-modelling the situation and asking all techniques to conform to the same modelling structure. Each scenario was thus modelled at an intermediate level of behavioural description. Also, fault trees and event tree based scenarios were used, so as not to favour any particular PRA representational approach. None of the assessors reported being particularly constrained by this pre-modelling, although one of the experienced THERP assessors did slightly remodel the scenarios, finding it necessary to go into more detail.

It should be noted that the pre-modelling was independent of the modelling using the HED method, which was generally far more global, with the exception of scenarios 2 and 5, which were modelled the same for HED and other techniques. Thus for three of the scenarios there would be no direct comparison between the various technique HEP's and the HED HEPs, but what would be possible would be a comparison between the total scenario HEPs (ie, the summation of each fault or event tree) for each technique, including the HED method. If these were found to be equivalent, this might suggest that detailed modelling was not always cost-effective in terms of accuracy, which would be an important finding.

Limitations of the Experiment

The experiment described in detail in the next section was one company's attempt to answer a fundamental question of human reliability assessment. Naturally, unlimited resources could not be spent on this experiment, leading to the following limitations: First, the data used to validate the techniques would ideally be wholly based on industrial applications and observations, but there were still not sufficient such data reported in industry to generate a large enough data base, and hence the recourse to human performance literature. Secondly, the need to co-locate experts for SLIM and APJ sessions proved sufficiently difficult that each expert group only met for one day, and carried out as many assessments as were possible in that period, but did not finish the entire set of scenarios. These were the two major limitations of the experiment.

Scenario Set

Five scenarios were abstracted from BNFL safety cases either recently or currently being completed using the HED method (HED results were not disclosed to any assessor/expert). The scenarios were as follows:-

Scenario 1 (S1): Manual Emergency Shutdown (ESD) occurs, followed by partial ESD failure, and subsequent need to detect, diagnose, and correct the failure, by recourse to procedural information and operations out on plant (see Figure 1).

Scenario 2 (S2): Calibration of monitors is carried out by plant personnel using a computerised system.

Scenario 3 (S3): Following a common cause fan failure, the operator has a short time to correct the situation, and there is potential for misdiagnosis and for using an innovative procedure for solving the problem.

Scenario 4 (S4):   During shutdown, the plant is washed. Areas of plant must be washed out consecutively and not in parallel. Administrative controls are the main features maintaining safety on this scenario.

Scenario 5 (S5a, S5b, S5c):   Three scenarios were modelled each concerned with an operator requesting a sample to be taken from a tank, then having it analysed, and acting on the results, over-seen by a supervisor and possibly also a plant manager.

Each scenario contained between 4 and 10 HEPs requiring quantification.

It should be stressed that the failure probabilities for these scenarios do not lead to the unwanted 'top event' associated with the safety case from which they have been abstracted since, as with most PRA's, many other hardware failures have to occur simultaneously in order to yield the top event. For reasons of confidentiality however, the assessors/experts were only shown the human error fault/event trees, and were not shown how they fitted into the much larger safety case fault trees. Any relevant information, eg concerning concurrent hardware failures which could impact on human performance, was given to the assessors, but not in the form of event or fault trees.

A detailed description of each scenario, plus a human operator fault/event tree, and a listing of the HEPs to be quantified, was given to each person taking part in the experiment (whether expert, assessor, or facilitator). Also there were handouts describing the plant control room layout, general aspects of the plant-operator interface, and pictures of the VDU system and typical VDU mimic diagrams, and an ESD panel. These general descriptions and pictures were generally supplied to participants two weeks in advance of the assessment.

<u>EXPERIMENTAL PROCEDURE</u>

The five techniques were run independently, as described below, so that each participant only participated in one technique, with the exception of Paired Comparisons (see PC below). The author attended all the expert group sessions solely as an observer and to clarify questions about the data set (containing known data points) and scenarios, where clarification was possible (this clarification function was also available to non-expert group techniques, ie to individual assessors carrying out THERP and HEART analysis).

The overall procedure was as follows:

THERP   –   three individual assessors worked independently, two of whom were recognised THERP assessors, the third being an individual who had not seen THERP before and was given ten days to become familiar with THERP, and some limited tuition in the technique. Each assessor was given the five scenarios and the data set to quantify, with no time limitations. All three assessors have a minimum of several years of safety assessment experience.

HEART	two ergonomists, one familiar with HEART, the other not, were asked to assess the scenarios and data set, again with no limitations on time. The ergonomist unfamiliar with HEART was given two hours tuition in the application of the technique by the technique's author. A third individual, a safety assessor, who was neither an ergonomist nor familiar with HEART, was given documentation on the technique and a few days to become familiar with it (it does not require as long as THERP), and then asked to quantify the scenarios and the data set.
APJ	two independent groups assembled each for a day, having familiarised themselves with the scenarios prior to the group session. Each group had a 'facilitator' to run the group, the facilitators both having previous experience of running APJ groups. Each group contained an ergonomist, two members with plant operational experience (5-35 years plant experience), and one member who was a safety assessor who had knowledge of the plant to which the scenarios related. Each group met for a day, and the facilitator led the group through a subset of the scenarios and the data set. The group was asked to reach consensus on each HEP where possible. Each group was in session for a full day (8-9 hours).
SLIM-MAUD	There were two SLIM-MAUD sessions , each with the same group personnel structure as for the APJ sessions (ie, one ergonomist, two personnel with plant experience, 10-35 years, and one safety assessor with knowledge of the plant whose scenarios were being assessed). The facilitator in each SLIM-MAUD group was highly familiar with the technique, and had run SLIM-MAUD sessions previous to the experiment. In one of the sessions, unfortunately, there was a 'bug' in the computer software which prevented the computerised version from running. This meant that the session had to be run on paper, which has the disadvantage of not being able to use the MAUD software to avoid expert judgement elicitation biases. However, since the group had assembled, it carried on with the assessments in this form. The results of the group session were later fed through another version of the program to generate the HEPs. The software worked in the other session.
Paired Comparisons	PC can be carried out as a questionnaire technique, providing that the questionnaire provides sufficient detail on the scenarios. Due to the high level of discussion that had taken place in the group sessions, it would have been difficult to document all of this in a form amenable to PC analysis. Therefore, experts who had participated in the panel session, and who were therefore familiar with the scenarios, were asked to complete a PC questionnaire

dealing with a subset of the scenarios and data set. This occurred approximately two months after the expert sessions, so that whilst experts were unlikely to remember the values assigned to HEPs in the group sessions, they would still remember the scenarios. Of the sixteen PC questionnaires sent out, nine were returned (completed) prior to the analysis of results taking place.

HED
Method   —   The HED approach had already quantified four of the scenarios prior to the experiment, and an assessor was asked to use the method to quantify S2 using the modelling structure defined for the experiment. Two assessors were also asked to use the method to quantify the 21 data points in the data set.

This was the overall design of the experiment, yielding 3 sets of THERP and HEART results, two sets of APJ and SLIM-MAUD results, and one set of PC and BNFL results, although only THERP, HEART, and BNFL approaches (for pragmatic reasons) finished assessing all five scenarios and the 21 data points. The following section discusses the results of the analysis, as analysed to date.

RESULTS

The following results are to an extent interim, since the analysis is still ongoing at the time of writing this paper. However, it is unlikely that the major conclusions of this paper will change, although others may be added.

Accuracy

Existence of a Predictive Relationship: Figures 2 to 7 show the results of each technique's predictions of a maximum of 21 data points (SLIM and PC did not estimate the total set). Taking logarithms of the estimates (to transform the data) and the data points, a correlation can be calculated (Pearson Product Moment Correlation Coefficient). The significant correlations found, in order of significance, were as follows (Note: Neither SLIM nor PC really produced enough estimates to satisfactorily yield a robust correlation coefficient):

	Correlation Coefficient (N = 21)	
HED 1	0.7705	($p <$ .000041)
APJ 2	0.7340	
HED 2	0.6958	
HEART 2	0.6553	
APJ 1	0.6446	
THERP 3	0.6311	
THERP 1	0.6095	($p <$ 0.0034)

Other correlations (HEART 1 and 3, and THERP 2) were not significant. It is interesting to note that HEART 2, which was significant, was the ergonomist with tuition on the technique, and that THERP 2, which was not significant, was the novice THERP assessor. Both of these results suggest that training may be required to use these two techniques (supporting the eighth hypothesis). Otherwise, the results suggest that each of these four techniques has some degree of empirical validity, and that the fifth hypothesis, that a significant relationship exists between some techniques and observed data, has been verified.

SLIM-MAUD and Paired Comparisons, due to resource limitations in the experiment , did not produce results which are properly amenable to this type of calculation (a correlation calculated for PC was not significant). However, a measure of precision can be calculated, as described below. A problem with the calibration of both of these techniques was noted, in that some extremely low HEPs were generated (eg, less than 10^{-10}) directly due to the experts failing to discriminate adequately between the two calibration points used in the experiment.

Precision: The correlation coefficient, if significant, shows that there is a relationship between the techniques estimates and the data base estimates. However, what is of practical interest in the PRA context is whether the estimates are in the right order of magnitude, ie whether they are correct to within a factor of ten of the 'real' value. On each of the graphs (Figures 2 to 7) two diagonal lines have been drawn at either side of the central diagonal which represents 'perfect agreement', this band marking what is tolerably precise in risk assessment terms (ideally of course, perfect agreement is desirable in which case all points would lie on the central diagonal).

If precision is defined as the proportion of HEPs within this band, then the following estimates of precision are derived from the graphs, and the number of optimistic and pessimistic (conservative) estimates are noted.

TABLE 1

	Total HEPs Assessed	Optimistic HEPs	Pessimistic HEPs	Precision %
HEART	63 (3 assessors)	8 (13%)	18 (29%)	59%
THERP	63 (3 assessors)	12 (19%)	14 (22%)	59%
APJ	42 (2 groups)	2 (5%)	8 (19%)	76%
SLIM	*11 (2 groups)	2 (18%)	3 (27%)	55%*
PC	*6 (9 assessors)	0 (0%)	2 (33%)	66%*
HED	42 (2 assessors)	7 (16%)	3 (7%)	77%

* With this low number of HEPs assessed, this measure of precision is unreliable

In terms of precision therefore, all the techniques offer between 55 and 80% precision, (although for SLIM and PC the total number of HEPs, is too small to give reliable results), with APJ and the HED method claiming the highest scores and, since APJ is less optimistic than the HED method (ie, it is more conservative), this measure would suggest APJ to be highest on this single criterion. Given that this level of precision is acceptable, this result verifies the first hypothesis, that some techniques will produce reasonably precise (accurate) predictions of observed data.

Convergence and Consistency: Inter-technique agreement (correlation of different techniques' results) could only be measured on some of the scenarios since APJ, SLIM, and PC did not finish all five scenarios.

All the techniques, with the exception of SLIM, correlated at least once with another assessor's/group's estimates using the same technique, showing a degree of consistency with these approaches (Note: PC consistency is tested in another way, and appropriate calculations have shown consistency within the PC judges). This shows a degree of consistency in the use of some of the techniques (verifying the second hypothesis).

The following statements summarise how techniques converged on the individual HEPs for at least one or more scenarios, or on the global results for one or more scenarios:

HEART correlated with APJ, SLIM, THERP, PC, HED
APJ   correlated with SLIM, THERP, HEART, PC but not HED
THERP correlated with HEART, APJ, SLIM, HED but not PC
SLIM correlated with HEART, PC, THERP, APJ, but not HED
PC   correlated with HEART, APJ, SLIM, but not THERP or HED
HED   correlated with HEART, THERP, but not PC, APJ or SLIM

HEART and THERP exhibited most correlations, although this is partly because they assessed all scenarios and thus had a greater opportunity to correlate. Scenarios 1 and 5 elicited most convergence from the techniques. Overall, there is a degree of convergence shown by the different approaches. From these results, HEART, APJ and THERP appear to have reasonable convergent validity, whereas the HED method only converged with THERP (however HED was only used decompositionally on one scenario in the experiment, a scenario in which no other techniques converged). SLIM and PC are more difficult to assess since whilst their uncalibrated scale values converged with some other techniques, the actual HEPs produced by these techniques did not fare so well. This again suggests that the calibration of SLIM and PC requires further development.

The global assessments (S1-S5c) could only be assessed for the three techniques (HEART, THERP, HED) which finished all the scenarios. It was interesting that whilst HEART and THERP agreed only partially on S1 and S5, nevertheless all HEART-THERP global comparisons correlated significantly, suggesting that paradoxically these two techniques overall tend to produce similar global results, even if the individual estimates within the event and fault trees are different. Also, the HED technique did produce global results which correlated significantly with at least one of the techniques, THERP. The global estimates are shown in Figure 8, and in this instance the HED estimates are used as a benchmark, (although this does not mean that these are necessarily the correct or validated HEPs). However, with the exception of one scenario, it can be seen that all the techniques generally agree with HED's estimates within the bounds of acceptable precision. Furthermore, most of the techniques' summated estimates for each scenario are within a band of two orders of magnitude or less, thus again suggesting general convergence of the techniques. Overall, convergence between certain techniques has been demonstrated by the results (verifying the third hypothesis).

Of particular interest is the fact that the HED method, using very global modelling of the scenarios, has generally derived the same order of magnitude of the global scenario HEPs, as the other techniques have using more detailed modelling. This suggests that detailed modelling may not be necessary to gain reasonably accurate results, and it is recommended that this whole area of modelling and the use of fault/event trees for representing human errors in safety cases, be further researched.

Homogeneity: Certain 'validated' techniques correlated with other 'validated' techniques on the scenarios, namely:

APJ and THERP
THERP and HED

This suggests that the data-set used to validate techniques has some degree of homogeneity with the scenario HEPs, verifying the fourth hypothesis, suggesting that the data set itself was a valid means of validating techniques for use in PRA.

Summary on Accuracy

Table 3 summarises the accuracy findings for the six techniques. The consistency and convergence criteria evaluated as functions of the degree of intra-technique and inter-technique correlation observed.

TABLE 2 : SUMMARY OF ACCURACY FINDINGS

	Precision	Consistency	Convergence	Predictive Validity+
THERP	Moderate	High	High	High
HEART	Moderate	Moderate	High	Moderate
APJ	High	Moderate	High/Moderate	High
SLIM	Moderate	Low	Moderate*	Undetermined++
PC	Moderate	High	Moderate*	Undetermined++
BNFL	High	Undetermined**	Low	High

* The uncalibrated PC and SLIM converged better than the actual HEPs
**The structure of the HED approach is designed to make it consistently used for safety cases, although this was not experimentally assessed.
+ Predictive validity means that there is a robust relationship demonstrable between the technique's predictions and actual values, even if these predictions do not precisely agree. Thus, for example, the PC and SLIM techniques' scale values fared better in this experiment than the calibrated HEPs did.
++Undetermined since not enough HEPs in the data set were estimated to reliably answer on this criterion.

Usefulness

In the introduction it was noted that HEART and SLIM are potentially highly useful techniques in aiding the assessor in determining how to cost-effectively reduce human error if required. In the experiment, neither PC nor the HED method produced any particularly useful insights in this way, as might have been predicted. Interestingly, THERP and APJ, which are usually seen as not especially 'useful', did produce this type of information (verifying the sixth hypothesis qualitatively). One THERP assessor produced a set of useful procedural information on how to improve the situation. The APJ sessions using expert groups, however, generated a surprising amount of useful information on risk reduction and safe practices which could be adopted (this also occurred in the SLIM sessions as expected). Whilst this information was not linked quantitatively to the HEPs, (so that a sensitivity analysis could not be easily carried out to determine the cost-effectiveness of the recommendations produced), THERP and APJ were nonetheless worthy of note as being of potentially significant qualitative usefulness. The techniques are therefore evaluated on this criterion as follows:

Most useful   —   HEART*
                    SLIM*
                    APJ
                    THERP
Least useful —   PC – HED

* Based on analysis from Reference 1.

It is hoped that further analysis of the results will produce some insights on the performance of the techniques with respect to the criterion of comprehensiveness.

Resources

The techniques are rated below on this criterion simply as a function of the person-hours the technique utilised to carry out scenarios (this is calculated pro-rata for those techniques which did not complete all scenarios). Other resources costs, such as software (eg, for SLIM), are not included, since for any company this would represent a one-off payment and would be in any case far less significant than the personnel time involved in sessions over time.

	Technique	Total/Projected Time for all Scenarios (one assessor or group)
Least resources	HED	2 days
	HEART	3-5 days
	PC	10 days
	THERP	8-30 days
	APJ	15 days (five personnel)
Most resources	SLIM	20 days (five personnel)

It is difficult to determine PC's resource requirements, since the PC participants in this study benefitted each from a detailed discussion on the scenarios. The actual time it takes to fill out a questionnaire for all scenarios would have taken approximately one day, and with preferably at least ten subjects, this yields 10 days. There was also wide variance in the use of THERP in that some very detailed analysis was produced by one assessor in particular (who incidentally was relatively accurate). SLIM was marginally slower than APJ. Additional weighting may be given to the fact that APJ and SLIM require the identification and co-location of a panel of experts. SLIM and PC also, in theory, require that calibration data is available with which to generate absolute HEPs.

Overall, resources requirements span an order of magnitude, demonstrating that techniques do differ in resources usage (hypothesis seven). This should, however, be considered by individuals selecting techniques with respect to other criteria, particularly usefulness, to determine the resource-effectiveness of techniques.

CONCLUSIONS AND RECOMMENDATIONS

1) THERP and APJ were found to be the most accurate techniques.

2) HEART showed promise on the accuracy dimension, and it is suggested that its development focus on achieving a higher degree of inter-assessor consistency.

3) The HED method was shown to be reasonably accurate.

4) SLIM and PC require a more robust calibration procedure to make the HEPs generated at least as valid as the scale values would appear to be.

5) The application of HEART and THERP both appear to benefit from the use of trained assessors in order to achieve the best results.

6) APJ and, to a lesser extent THERP, were both found to produce useful qualitative information, although this was not quantitatively linked to the HEPs as for SLIM or HEART (the most useful techniques).

7) In terms of resources, HEART (and HED) required least resources, with APJ and SLIM requiring most. THERP and PC required a moderate degree of resources, with the degree of THERP resources required being fairly variable, dependent upon the individual assessor.

8) Further analysis of the results will concentrate on investigating the performance of the techniques with respect to the criterion of comprehensiveness, although a definitive evaluation against this criterion would ideally require a further experiment.

9) The area of quantification is critically dependent on the validity of the modelling of scenarios, and it is recommended that more attention be switched from pure quantification to error identification and representation issues. In particular the results of this experiment raise the important question of whether detailed modelling is necessary to gain accurate results.

ACKNOWLEDGEMENTS

The author wishes to thank George Hensley for his inspiration and support of this work within BNFL Safety Department; Jerry Williams (CEGB) for his suggestions on the experimental design and the generation of usable data; Julie Reed (Electrowatt), Andy Smith (Atkins), and Richard Sharp (HELP) for assistance on the analysis of results; Barry Whittingham (Electrowatt) for his assistance on the development of the scenarios and as an assessor; Keith Rea (NNC), Ned Hickling, Trevor Waters, Joanne Pennington and Bill Gall (SRD), Sue Whalley (Lihou Loss Prevention), David Embrey (HRA) and A Smith, B Whittingham and R Sharp for their role as assessors/facilitators in the experiment, and to the personnel at Risley and Sellafield who also participated in the experiment as assessors/experts: R Benson, H Chapman, J Foster, J Freestone, P Hancock, G B Guy, H Jones, N James, J Lodge, J Lomax, B Martin, P Redfern, J Rigg, H Rycraft, C Ryder, G Sheppard, K Waterhouse, and J Whitford.

REFERENCES

1) Human Reliability Assessment Group (in press) The Human Reliability Assessor's Guide. To be published by NCSR, UKAEA, Wigshaw Lane, Culcheth, Cheshire.

2) Williams, J C (1983) Validation of Human Reliability Assessment Techniques. In Proceedings of the Fourth National Reliability Conference, 6-8 July, NEC Birmingham. Published by NCSR and IQA.

3) Swain, A D and Guttmann, H E (1983) A Handbook of Human Reliability Analysis with Emphasis on Nuclear Power Plant Applications. Nureg/CR-1278, USNRC, Washington DC-20555.

4) Embrey, D E (1982) Can We Predict Human Error in High Risk Technology. Paper presented at the Annual Conference of the British Association for the Advancement of Science, Liverpool, 10 September 1982.

5) Embrey, D E, Humphreys, P C, Rosa, E A, Kirwan, B, and Rea, K (1984) SLIM-MAUD: An Appraisal to Assessing Human Error Probabilities Using Expert Judgement. Nureg/CR-3518, USNRC, Washington DC-20555.

6) Kahneman, D, and Tversky, A (1979) Intuitive Prediction: Biases and Corrective Procedures. TIMS Studies in Management Sciences, 12, pp 313-327. Weelwright and Maridatis, eds.

7) Pontecorvo, A B (1965) A Method of Predicting Human Reliability. Annals of Reliability and Maintenance, 4.337-342. Fourth Annual Reliability and Maintainability Conference.

8) Hunns, D M (1982) The Method of Paired Comparisons. In: Green, A E (Ed), High Risk Safety Technology. Wiley: Chichester.

9) Seaver, D A, and Stillwell, W G (1983) Procedures for Using Expert Judgement to Estimate Human Error Probabilities in Nuclear Power Plants Operations. Nureg/CR-2743, USNRC, Washington DC-20555.

10) Williams, J C (1986) HEART - A Proposed Method for Assessing and Reducing Human Error. In Proceedings of the 9th Advances in Reliability Technology Symposium, University of Bradford, 4 April. Published by NCSR, UKAEA, Culcheth, Cheshire.

11) Rasmussen, N (1985) Reactor Safety Study Atomic Energy Commission Report WASH-1400, Washington DC.

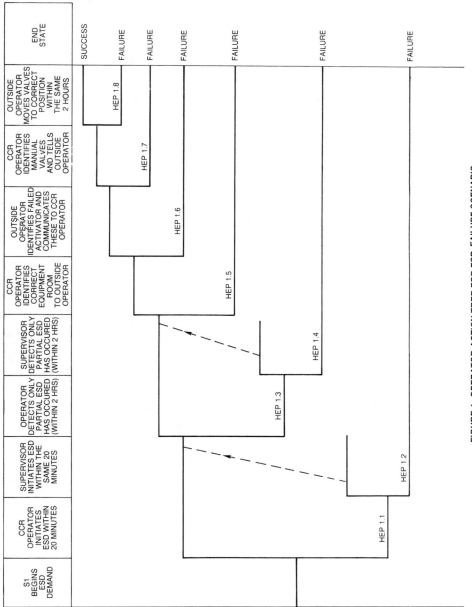

FIGURE 1: OPERATOR ACTION TREE FOR ESD FAILURE SCENARIO

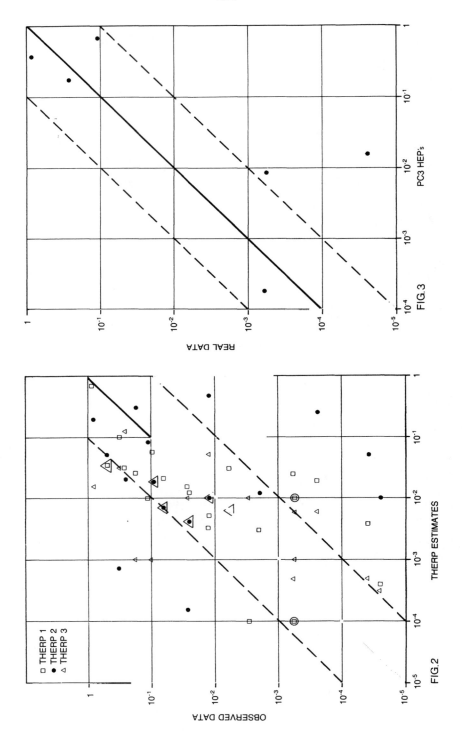

FIG.2

FIG.3

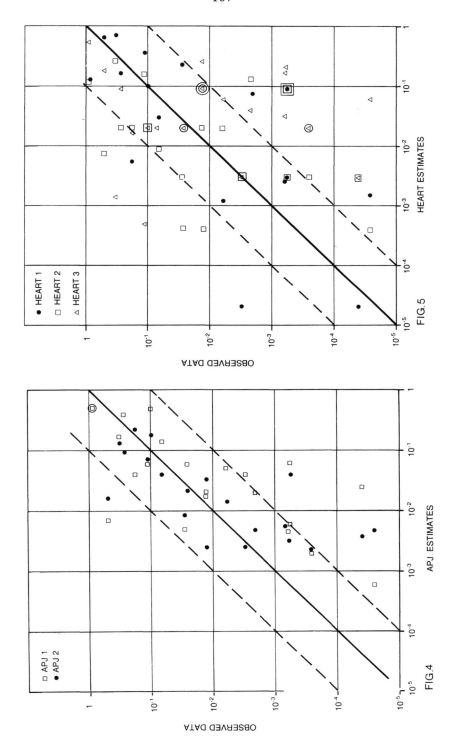

FIG.4

FIG.5

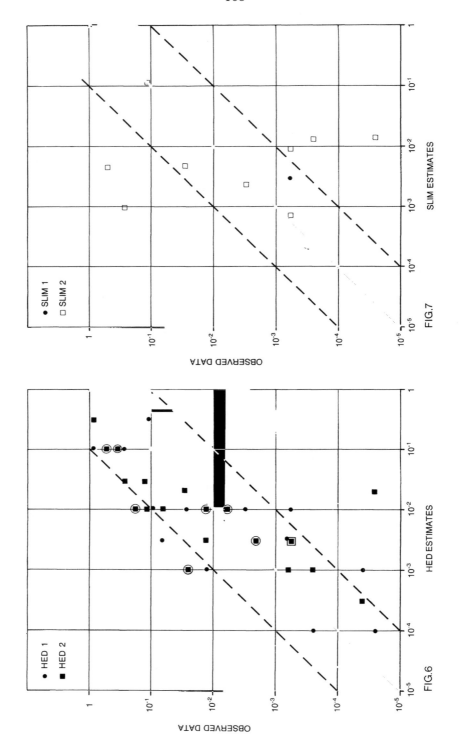

FIG.6

FIG.7

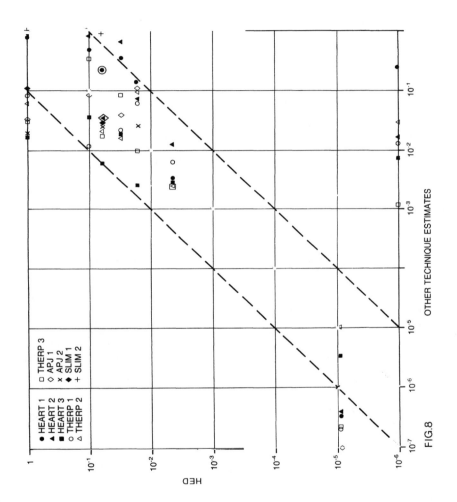

FIG.8

110

HUMAN FACTORS RELIABILITY BENCHMARK EXERCISE,
REPORT OF THE SRD PARTICIPATION*

TREVOR WATERS
Reliability Technology Research Unit
National Centre of Systems Reliability,
Safety and Reliability Directorate, UKAEA
Wigshaw Lane, Culcheth, Warrington WA3 4NE

ABSTRACT

Within the scope of the Human Factors Reliability Benchmark Exercise,
organised by the Joint Research Centre, Ispra, Italy, the SRD team has
performed analysis of human factors in two different activities – a
routine test and a non-routine operational transient.

For both activities, an 'FMEA-like' task analysis has been performed
to identify the requisite tasks, potential errors, and the factors which
affect performance. For analysis of the non-routine activity, which
involved a significant amount of cognitive processing, such as diagnosis
and decision making, a new approach for qualitative analysis has been
developed. Modelling has been performed using both event trees and fault
trees and examples are provided. Human error probabilities were
estimated using the methods APJ, HCR, HEART, SLIM, TESEO, and THERP. A
discussion is provided of the lessons learnt in the course of the
exercise and unresolved difficulties in the assessment of human
reliability.

INTRODUCTION

The Human Factors Reliability Benchmark Exercise (HF-RBE) is the third of
four benchmark exercises organised by the Joint Research Centre, Ispra,
Italy. The series began with the Systems Reliability Benchmark Exercise,
which was followed by the Common Cause Failure Reliability Benchmark
Exercise. The series of benchmarks will be completed by the Event
Sequence Reliability Benchmark Exercise which serves to draw together the
experience gained over the series into a single analysis.

*copyright retained by UKAEA

Aims
The aims of the HF-RBE are:

- to assess the degree of consistency of methods and results of human reliability analysis
- to identify consensus analysis procedures

Scope
The system which was analysed for the HF-RBE was the Kraftwerk Union (KWU) Pressurised Water Reactor at Grohnde, and in particular, analysis concentrated on the Emergency Feed System (EFS) which was the basis of analysis of the previous RBEs.

One of the functions of the EFS is to remove decay heat in the event of loss of feedwater to the steam generator (SG).

Plant safety depends upon the timely operation of the EFS which is itself dependent upon two types of operator activity:

- routine testing of components
- fault management

Both types of activity were analysed separately during the course of the exercise, their characteristics differ markedly and will be described later.

Outline of Exercise
A total of fifteen teams from throughout Europe and USA took part in the analysis which was spread over a period of two years, between April 1986 and March 1988.

The following general procedure was followed for the analysis of both activities:

- examination of the documentation provided (this consisted of translated excerpts from the written procedures and a video recording of the operator's activity)
- clarification of technical details (via submission of questions to KWU)
- analysis
- evaluation of the analysis at the subsequent meeting of participants and, if necessary, modification of the scope

A summary of the contributions from the different participants can be found in reference [1].

112

STUDY CASE 1 – ROUTINE TEST ACTIVITY

The testing of components and systems clearly has an important effect upon overall reliability. Behaviour in such situations tends to be characterised by the use of written procedures, in addition to which the situation is familiar, stress levels are low and 'lower' levels of cognitive processing are involved. The test procedure under analysis concerned the functioning of the emergency feed system, as described in the previous section.

Phase 1 – Initial Analysis
In this initial analysis the test procedures investigated were:

- the cut-in signal (which opened an isolation valve and the SG control valve)
- the shut-off signal (which closed these valves)
- the functional test of components in one train of the EFS

Because the cut-in and shut-off signal tests were related they were considered as one task. This task and the functional test of components were examined in relation to three failure conditions:

- unavailability of the EFS due to human error during the tests
- failure of the operator to detect a failed component
- initiation of an operational transient by human error during the tests

Preliminary analysis indicated that on the basis of the information provided, it was not possible to determine the third failure condition (initiation of an operational transient). The remainder of the section describes the analysis of the two tests (cut-in/shut-off and test of components) for the two failure conditions (unavailability due to human error and failure to detect a failed component).

Test of Cut-in and Shut-off Signals: To perform this test the operator uses written procedures but is also guided by an automated system on computer. If errors are made during the test then the computer prevents progression of the procedure. As a result no errors are possible other than total omission of the test. This 'administrative' type omission error was modelled using fault trees and quantification made using the techniques: THERP, HEART and TESEO.

Functional Test of Components: During the test procedure the following components are tested:

Emergency feed pump
Free flow check valve
Demineralised water circulation pump
Two way valve
Flow limitation valve

Figure 1 shows an excerpt from the test procedure, showing the steps required to test the functioning of the free flow check valve. The operator errors were modelled using both fault trees and event trees and quantification was done using APJ, HEART, TESEO and THERP.

113

```
┌─────────────────────────────────────────────────────────────────────────┐
│ Components of the Emergency Feedwater Subsystem RS10                       │
│ Routine Functional Test                                                    │
│                                                                            │
│     Test free flow check valve RS12 S001                                   │
│                                                                            │
│ (5)  - FLOW EF-PUMP AT MINIMUM FLOW            RECORD       0 RS13 F501     │
│        ┌────┬────┬────┬────┐                   Setpoint                     │
│        └────┴────┴────┴────┘                   6,1-6,5 kg/s                 │
│                                                                            │
│      - Actual value corresponds to setpoint                                │
│                                                                            │
│      - FLOW EMERGENCY FEED PUMP                RECORD       0 RS12 F001     │
│        ┌────┬────┬────┬────┐                   Setpoint                     │
│        └────┴────┴────┴────┘                   0 kg/s                       │
│                                                                            │
│      - Actual value corresponds to setpoint                                │
│                                                                            │
│      - VENTING VALVE UPSTREAM    RS12 S002     OPEN         0 RS12 S801     │
│      - VENTING VALVE DOWNSTREAM RS12 S801      OPEN(SLOWLY)/0 RS12 S802     │
│        (Opening of the valve up to the        CLOSE                        │
│        closure of the side-port of the free                                │
│        flow check valve,                                                   │
│        Flow side-port 0 kg/s indicated by                                  │
│        RS13 F501;                                                          │
│        RS12 S802 then has to close immediately)                            │
│      - VENTING VALVE UPSTREAM    RS12 S002     CLOSE        0 RS12 S801     │
└─────────────────────────────────────────────────────────────────────────┘
```

Figure 1. Excerpt from the test procedure

   Evaluation of Phase 1: Useful experience was gained in the problems
of modelling and quantification during this first phase, and these are
included in the general discussion at the end of the report. A direct
comparison of the numerical results between different teams is difficult
due to the differences in assumptions made by different participants.
The different assumptions concern:

   1   the functioning of the system, eg the criteria and symptoms of
       different component failures
   2   human error modelling, eg the role of the supervisor
   3   the application of quantification methods

   To facilitate comparison of results it was agreed to perform a
second phase in which the scope of the analysis was limited, this is
described in the following section.

Phase 2 - Limited Scope Analysis

In this limited scope analysis the two failure conditions investigated were:

- unavailability of the EFS due to the operator leaving two venting valves open during the functional test of components
- failure to detect the failed free flow check valve (in three possible failure modes)

Failure to Detect the Failed Free Flow Check Valve: This valve has a single inlet (from the emergency feed pump) and two outlets (one main port to the steam generator and one side port to the demineralised water storage tank). The valve is designed to ensure that if feed is not required to the steam generator then the side port will open to enable the EF pump to operate under minimum flow conditions. (The valve also protects the EF pump against back flow when the system is not in operation.)

Figure 1 shows the procedure for the test of this component. With the EF pump running, and the isolation valve to the SG shut, the flow rate in the side port and main port are recorded in the written procedure. The venting valves in the main port line are then opened, this should cause the main port to open and the side port to close, which is indicated by the flow in the side port reducing to zero. The test is completed by closing the venting valves.

Three component failure modes were chosen:

- the side port remains open (interpreted as meaning: the side port does not close fully)
- the side port fails to open when required
- the main port fails to open when required

It was assumed that the initial standby state of the free flow check valve (FFCV) was with the main port closed and the side port fully open. Another important assumption was that the failure occurred in the period between the end of the previous monthly minimum flow test and the start of the current test.

Each failure mode was modelled using an event tree, which was then quantified using THERP. Quantification was also performed using HEART, but with no decomposition, ie each task was considered as a whole. An attempt was also made to quantify the failure modes using SLIM, seven PSFs were identified and the individual tasks categorised. Unfortunately, no suitable reference data was available and so quantification was prevented.

Unavailability due to the Operator leaving Venting Valves Open: The venting valves are opened twice during the monthly test of components, once during the test of the free flow check valve described earlier and once at the end of the procedure to relieve pressure in the main line. These tasks were assumed to be independent and it was considered that errors made on the first venting would be recovered.

The event tree produced to model this error is shown in Figure 2. Quantification was made using THERP (decompositionally) and HEART (holistically).

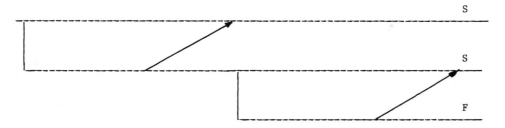

Figure 2. Event tree to model the scenario in which the venting valves are left open

Evaluation of Phase 2: The results of quantification of the different errors are shown in Table 1.

It can be seen that the two methods of quantification agree roughly with the ranking of human errors. It was generally the case that THERP produced lower error probabilities than HEART, though without actual field data it is not possible to assess which method is more accurate.

TABLE 1
Comparison of quantitative results for phase 2

Human Error	Quantification Method	
	HEART	THERP
FM1 Fail to detect that side port remains open	0.02	0.006
FM2 Fail to detect that side port does not open when required	0.1	1.0
FM3 Fail to detect that main port does not open when required	0.02	0.004
Venting valves left open after test	0.003	0.0002

Phase 3 - Analysis using Standardised Failure Mechanisms
In order to further facilitate comparison of methods of quantification it was decided to restrict the scope of the analysis again, to a single failure of the FFCV, namely, side port fails to close, which is identical to FM3 described above.

The four different error modes which were agreed were:

a   Operator 1: omission of steps 5.5 and 5.6 in the procedure
b   Operator 2: fails to recover omission
c   Operator 1: omission to read flow display
d   Operator 1: misread the flow display

The event tree produced is shown in Figure 3. Table 2 shows how the individual error probabilities were calculated using THERP. It was considered that task c was totally dependent upon tasks a and b. That is, if the venting valves were opened then the display would be read. The calculated human error probability was the same as the previous phase, ie 0.004.

Using HEART generic task type 'E' was considered most appropriate, ie "Routine, highly practised rapid task involving relatively low level of skill", nominal human error probability = 0.02. There was not considered to be any significant error producing conditions and so the error probability was not modified.

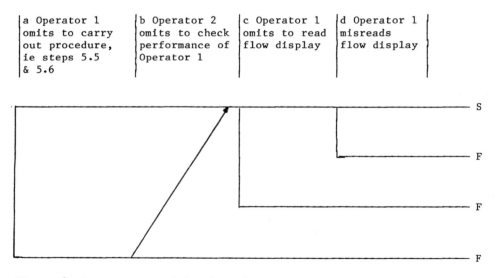

Figure 3. Event tree used in phase 3

TABLE 2
Derivation of quantitative information using THERP

Description	THERP Data Table	Nominal HEP (Item)	Error Factor	Basic HEP	Dependency	Conditional HEP
a Operator 1 omits to carry out procedure, ie steps 5.5 & 5.6	20:7	0.003 (2)	3	0.001[1]	Not applicable[2]	Not applicable[2]
b Operator 2 omits to check performance of Operator 1	20:22	0.2 (2)	5	0.9[3]	Not applicable[4]	Not applicable[4]
c Operator 1 omits to read flow display	20:7	0.003 (2)	3	0.001[1]	CD[5]	0
d Operator 1 misreads flow display	20:11	0.003 (4)	·3	0.003	ZD[6]	0.003

[1] Although the full procedure is long and contains more than ten steps, it is divided into clearly defined stages corresponding to the test of specific components. Therefore the Nominal Human Error Probability (HEP) was reduced by the full Error Factor.
[2] Dependency and hence Conditional HEP is not applicable for the first task on the event tree.
[3] It would appear that Operator 2 is not actually responsible for checking the performance of Operator 1. Although there is some possibility that the operator will notice the omission, to reflect the reality of the situation, the HEP has been increased accordingly.
[4] Dependency between operators is included within the HEP.
[5] See text.
[6] Success at step a and b is not considered likely to affect the probability of success at this task.

STUDY CASE 2 – OPERATIONAL TRANSIENT

The second study case concerned operator intervention during an abnormal event. The situation is different from the routine test activity for a number of reasons, for instance the level of stress is likely to be higher, time constraints are stricter, the situation is less familiar, the written procedures are more complex and higher levels of cognitive processing are involved, particularly activities such as diagnosis and decision making. It was intended to investigate how these factors affect both the analytical approach and the level of operator performance.

Description of the Event
The event starts with a loss of electrical supplies which causes a reactor trip, and actuation of four emergency diesels. However, two

diesels (which power the shutdown pumps) fail to start and there is no feed to the SG. It is assumed, for the purpose of the exercise, that the failed diesels cannot be repaired. Levels in the SG fall to 'low low' causing cut-in of the emergency feed system. The event is further complicated by a common mode failure of the isolation valves in each of the four trains meaning that SG feed is still insufficient. The operator is required to diagnose valve failure and open the valves manually within about forty minutes of cut-in of the EFS.

Qualitative Analysis

A qualitative analysis was performed which included a decompositional 'FMEA-like' task analysis, an extract from which is shown in Table 3. A method was also developed in an attempt to systematise the identification of cognitive errors and causes of errors in complex situations. This method, called the Critical Actions and Decisions Approach (CADA) has been developed from the work done by Pew [2], in which actions and decisions are considered in eight stages:

    Detection of signal
    Observation/data collection/investigation
    Identification of state of plant
    Interpretation of situation/understand implications
    Evaluation or review of alternative strategies/objectives
    Select objective/goal state
    Select procedure
    Execute procedure

An error made at any stage can cause an inappropriate action. The analyst selects the stages that are relevant, and subsequently administers a checklist of questions to identify potential causes of error. The procedure is intended to flag up potential errors and indicate factors which are known to affect the operator's cognitive performance, eg the tendency for people to strive to confirm an existing hypothesis and reject or rationalise conflicting data. The method is at an embryonic stage requiring considerable development and is described more fully in reference [3].

It was assumed that operators would follow the written procedures and as a result there was a high likelihood that the operators would initially interpret the situation correctly and anticipate the initiation of the emergency feed system. There is a small possibility that the operator could misinterpret the situation initially, in which case it could be postulated that in certain circumstances the 'hazard' alarm (which is annunciated when the 'low low level' is reached in the SG) could confuse or mislead the operator further. However it was considered that the 'hazard' alarm was sufficiently compelling and the training adequate such that the probability of occurrence of both errors would be small and would be dominated by the probability of error of subsequent tasks, particularly the cognitive task involving interpretation of the situation.

TABLE 3
Results of the qualitative analysis

Task	Errors	Consequences	Comments
REACTOR HAS TRIPPED AND IS OPERATING IN EMERGENCY POWER MODE, DIESELS 1 AND 2 HAVE FAILED.			
Identify that the SG level is falling	Failure to identify SG fall	Operator will not be antici-pating cut-in of emergency feed system. Subsequent stress may be higher.	On basis of infor-mation provided it was assumed that the operators would follow procedures. No other events had similar symptoms and misdiagnosis was con-sidered unlikely.
LOW-LOW LEVEL REACHED HERE, 'HAZARD' ALARMS ANNUNCIATED. EFS CUTS IN.			
Check SG parameters in operating manual part 3-01 and iden-tify partial operation of EFS isolation valves	Failure to check any parameters. Fail to check parameters in correct sequence and terminate checking early.	Operator fails to identify that the emergency feed isolation valves are only partially open. Operator may diagnose that the emergency feed pumps are degraded.	There is no direct written instruction to the operator to check the SG parameters when the EFS is working normally. However, it is expected that the operator would monitor SG levels continuously and would detect the continuation of the SG level fall.
Locate plan and read valve location numbers	Fail to locate plan. Read incorrect valve location	Restorative action will be delayed, increased stress. APOs will be directed to wrong location, causing delay.	Individual reading errors are unlikely to be serious, as there are four valves of which only one needs to be opened. A dependent failure is possible if the shift supervisor reads the wrong valve system-atically, eg RS17-47 S004 instead of RS12-42 S004.
Send out call by loudspeaker to Auxiliary Plant Operators (APOs)	Fail to make call. Fail to operate micro-phone. Message is ambiguous.	APOs fail to phone control room.	Errors are difficult to predict with the information provided. Recovery is probable, main consequence will be time delays.

Modelling
The event was modelled using an event tree which is shown in Figure 4.

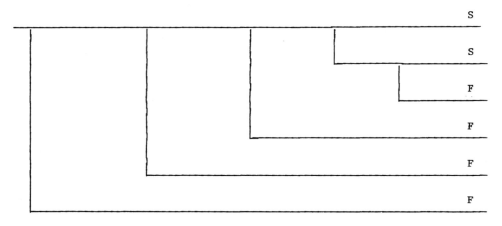

Figure 4. Event tree of operator intervention during the operational transient

Quantification
Four separate analyses (based on the event tree shown above) were performed using different combinations of HCR and THERP. In addition, for each analysis two calculations were performed, one using optimistic time assumptions and one using pessimistic time assumptions. The eight separate results are shown in Table 4. The different analyses were performed to investigate the effect of different assumptions (eg time taken by operators to perform different tasks) and also different event tree models. It can be seen that the most pessimistic analysis (A) is more than two orders of magnitude greater than the most optimistic analysis (D). The completeness and accuracy of the qualitative understanding clearly has a significant effect upon the quantification phase. On the basis of the information provided, the authors best point estimate is 0.05.

An attempt was made to quantify the errors using SLIM, a generic data base was used as the source of reference data with the PSFs and their weightings predefined. The calculated human error probability using SLIM was 0.4.

TABLE 4
Summary of results from four analyses (using optimistic and pessimistic time assumptions)

Analysis	Sub-Task	Estimated Nominal Median Time to Perform Sub-Task (minutes)	Calculated Conditional Human Error Probability	Method of Quanti-fication	Overall Human Error Probability
A Using HCR for·all sub-tasks	1	6	0.5	HCR	
	2	4	0.5	HCR	
	3	7	0.5	HCR	0.9
	4	2	0.5	HCR	
	5	2	0.5	HCR	
	1	4.5	0.14	HCR	
	2	2	0.1	HCR	
	3	2.5	0.1	HCR	0.3
	4	2	0.1	HCR	
	5	2	0.1	HCR	
B Using a combination of HCR and THERP	1	6	0.049	HCR	
	2	4	0.0095	THERP	
	3	7	0.03	THERP/APJ	0.06
	4	2	0.006	THERP	
	5	2	0.503	THERP	
	1	4.5	0.0028	HCR	
	2	2	0.0095	THERP	
	3	2.5	0.03	THERP/APJ	0.02
	4	2	0.006	THERP	
	5	2	0.503	THERP	
C Omitting to model errors in tasks 2 and 3	1	6	0.049	HCR	
	2	4	0	–	
	3	7	0	–	0.05
	4	2	0.006	THERP	
	5	2	0.503	THERP	
	1	4.5	0.0028	HCR	
	2	2	0	–	
	3	2.5	0	–	0.006
	4	2	0.006	THERP	
	5	2	0.503	THERP	
D Omitting to model errors in tasks 2 and 3 and with no considera-tion of time	1	6	0.003	HCR	
	2	0	0	–	
	3	0	0	–	0.006
	4	2	0.006	THERP	
	5	2	0.503	THERP	
taken to perform these	1	4.5	0.001	HCR	
	2	0	0	–	
	3	0	0	–	0.004
	4	2	0.006	THERP	
	5	2	0.503	THERP	

122

DISCUSSION

In the following sections a number of points of interest and implications arising from the work described earlier will be discussed. Three separate aspects will be covered to reflect the analytical stages which have been followed: qualitative analysis/task analysis, modelling/representation and quantitative assessment.

Qualitative Analysis/Task Analysis

A comprehensive and systematic qualitative analysis is essential if the assumptions upon which the remainder of the analysis is made are to be realistic. The purpose of the qualitative analysis is to:

1   Identify the tasks which are performed
2   Identify the factors which affect performance of the task
3   Identify the errors which could be made
4   Identify the consequences of the errors
5   Identify the root causes of the errors

A decompositional 'FMEA-like' task analysis format has been used throughout the HF-RBE. This is considered to be a generally useful and auditable method to achieve the aims described above, particularly for procedural tasks which consist of a well defined sequence of actions.

Clearly performing a detailed qualitative analysis on the basis of a video recording, a set of written procedures and supplementary technical information provided by KWU is not ideal, though it is realistic and probably more than would be available for the assessment of a new system. A certain amount of guesswork and inference was required in the analysis of the task, and many subtleties of task performance may have been unnoticed. For instance, auditory or visual signs and symptoms may be learnt and used by the operators although they may not be described in the written procedures. In addition, factors such as motivation and the operators perception of risk are difficult to assess without direct access to the operators. These limitations should be borne in mind when performing an analysis. Particularly for critical errors which impact heavily on the overall system reliability, then further information and contact with operators is desirable.

The prediction of errors is generally easier for routine procedural activities than non-routine activities involving cognitive tasks such as diagnosis and decision making where many different outcomes can occur. However, as experienced in phase 1 of the routine test analysis, the assessment of the probability of the operator causing an operational transient, by making a commission error, is less easy and requires more detailed system information. In study-case 1, errors were predicted without a formal structured framework and relied totally on the analysts experience. Techniques such as SHERPA [4] can be used to systematise the process of error analysis in proceduralised situations.

Error identification in non-routine, complex situations which involve high level cognitive processing is more difficult. Operator variability in such situations is likely to be greater, and the assumptions concerning operator behaviour (eg that written procedures will be followed) may be more difficult to justify. These factors generally result in greater levels of uncertainty which will be reflected in the modelling and quantification stages. For such types of behaviour interviews with operators, simulation exercises and study of past operating experience can be used to improve the analysis. The CADA method for the analysis of cognitive errors which was developed for this exercise, is intended to provide a more systematic method for the identification of potential errors and causes. However, it is obviously in an early stage and further development is required.

It is important that the task analysis is not constrained to the errors and PSFs which may be described within the quantification method. For example, THERP contains quantitative guidance for the effect of both stress and experience and also data for over 20 different classes of error. However there are many other PSFs which may have a significant effect in specific situations and many other types of error. Specific advice is provided by THERP to combat this difficulty. These extra considerations inevitably cause further work, but are considered necessary if the subsequent quantification is to be realistic.

Modelling

The purpose of modelling is to provide a structure for quantification, although it also provides useful qualitative insights. Both event trees and fault trees have been used in this analysis and choice can be made according to the preference of the analyst, as both should produce identical results. This author prefers the use of event trees particularly for modelling errors during activities which take place over a period of time. The timewise development of an event across the page reflects naturally the timewise development of an event or procedure.

Event trees tend to depict behaviour in terms of success or failure, however this is not always appropriate for many tasks such as diagnosis and decision making, which can have a large number of different outcomes. It is useful to state the various outcomes explicitly both for a qualitative understanding and to aid auditing.

Quantification

Six different methods of quantification have been applied during the course of the HF-RBE:

APJ [5]
HCR [6]
HEART [7]
SLIM [8]
TESEO [9]
THERP [10]

None of the methods are without problems, which is perhaps not surprising given the complexity of human behaviour. There would appear to be no single best method, but rather the most appropriate will depend upon the characteristics of the specific task under consideration. It is outside the scope of this document to provide comprehensive guidance for the evaluation of different methods. For further information about these methods the reader is referred to the paper by P Humphreys presented earlier in the symposium.

General Discussion

In both the routine test procedure and the non-routine emergency situation the qualitative analysis/task analysis is important if the assumptions upon which the quantitative analysis depend are to be realistic. This is generally easier for routine procedures where actions are well defined and familiar, although verification of the assumptions made may be useful for important tasks. For example, interview with operators may be required to verify that procedures are actually followed and that short cuts are not taken. In routine procedural situations, methods such as SHERPA can be useful for the systematic prediction of errors. Qualitative analysis in non-routine situations is less straightforward however, particularly where cognitive tasks such as diagnosis and decision making are involved, which can have a number of outcomes. The CADA method which has been developed within this HF-RBE is intended to facilitate this activity, though clearly further development is required. For these situations verification of predicted operator behaviour is particularly beneficial. Without such verification, the uncertainty involved in quantification will be greater. It was shown in study case 2 that differences in assumptions involving performance of the task and in the use of quantification methods resulted in overall difference of two orders of magnitude in the overall estimated human error probability.

Modelling or representation of human errors is an important stage both for a qualitative understanding and quantification. It is crucial that modelling is driven by the results of the qualitative analysis and not by the characteristics of the method for quantification.

There would currently appear to be no single method of quantification which is universally applicable, and a certain amount of 'intelligent' interpretation is required if the methods are to be applied appropriately. All of the methods have problems, however to assess the importance of these problems with respect to accuracy, it is necessary to perform further work to compare predicted human error rates against actual field data.

CONCLUSIONS

1   A detailed qualitative analysis is a prerequisite if the assumptions upon which quantitative analysis is made are to be realistic.

2   Methods such as SHERPA and CADA are useful tools for the prediction of errors and the identification of performance shaping factors, although further development is required.

3   For important tasks and errors, the results of qualitative analysis should be verified to ensure validity and completeness. This can be done by interview with operators, simulation etc and helps reduce the uncertainty of the final quantification.

4   Event tree modelling is generally preferred to fault trees for sequences of actions.

5   No single method for quantification is universally applicable. There are problems involved in the use of all of the methods and further validation work is required to evaluate the accuracy of predictions.

REFERENCES

1. Poucet, A., Human Factors Reliability Benchmark Exercise, Final Report. To be published as Euratom Report PER 1482/88.

2. Pew, R.W., Miller, D.C. and Feeher, C.E., Evaluation of proposed control room improvements through analysis of critical operator decisions. Research Reports Center, Box 504 90, Palo Alto, CA 94303, USA. EPRI Report NP-1982, 1981.

3. Waters, T.L. and Penington, J., SRD unpublished work.

4. Murgatroyd, R.A., Embrey, D.E., Ballard, G. and Tait, J., The reliability of ultrasonic inspection. In Proc 8th International Conference on non-destructive evaluation in the nuclear industry, Orlando, Florida, USA, 17-19 November, 1986.

5. Comer, M.K., Seaver, D.A., Stillwell, W.G. and Gaddy, C.D., Generating human reliability using expert judgement. NUREG/CR-3688, 1984.

6. Hannaman, G.W. and Worledge, D.H., Some developments in human reliability analysis approaches and tools. International Post-Smirt 9 seminar on accident sequence modelling: human actions, systems response, intelligent decision support, Munich, West Germany, 24-25 August, 1987.

7. Williams, J.C., HEART - a proposed method for assessing and reducing human error. Proceedings of 9th Advances in Reliability Technology Symposium, University of Bradford, England, 1986.

8. Embrey, D.E., Humphreys, P., Rosa, E.A., Kirwan, B. and Rea, K. SLIM-MAUD: An approach to assessing human error probabilities using structured expert judgement. NUREG/CR-3518, 1984.

9. Bello, G.C. and Colombari, V., The human factors in risk analysis of process plants: the control room operator model 'TESEO', Reliability Engineering, 1, 1980, p3-14.

10. Swain, A.G. and Guttmann, H.E., Handbook of human reliability analysis with emphasis on nuclear power plant applications. NUREG/CR-1278, 1983.

The Application of the Combined THERP/HCR Model
in Human Reliability Assessment

R.B. WHITTINGHAM
Electrowatt Engineering Services (UK) Ltd
Grandford House, 16 Carfax, Horsham,
West Sussex, RH12 1UP. England

ABSTRACT

The paper describes how two existing techniques of Human Reliability
Assessment have been combined to produce a flexible methodology for
modelling human error situations. THERP [1] is a well established data base
method which is used to model a wide variety of manual and visual tasks.
HCR (Human Cognitive Reliability) model [2] is a recent development for
modelling diagnostic and decision making responses under time restrictions.
By combining the two approaches, almost any sequence of human activities can
be modelled such that error tendencies are identified and quantified in
probabilistic terms.

INTRODUCTION

There is an increasing awareness among the owners and operators of hazardous
plant of the impact of the human being when attempting to achieve safety and
risk targets. Methods of probabilistic risk assessment, developed to a high
degree of acceptability over the last two decades, have come to be seen as
having limited value unless taking into account the inevitable human factor.
It is important therefore that developments in the quantification of human
reliability keep abreast of progress in the assessment of equipment
reliability.

This paper attempts to address this situation by describing a methodology
for Human Reliability Analysis (HRA) which, it is believed, can satisfy a
number of important criteria. It is suggested that any credible HRA
methodology must attempt to achieve the following:-

1. It must be comprehensive. Identification of all human error
   possibilities is essential if the exercise is to verify that safety
   targets are achievable. This is also the most difficult criteria in
   terms of methodology and proof.

2. It must be systematic. Ideally the methodology should be carried out
   within a framework which enables both the assessor and the client to
   draw valid conclusions from the result.

3. It must be scrutable. The methodology and modelling must be such that the reader is able to understand how the result has been achieved and be able to relate the model to the real world.

HUMAN RELIABILITY ANALYSIS

The need to predict human error probability (HEP) arises on both new and existing plant. It is undertaken usually where the consequences of human error are severe either in the matter of financial penalty or more often where there is a major hazard arising from plant operation. Human Reliability Analysis (HRA) on new plant is becoming increasingly necessary to meet more stringent licensing and regulatory requirements. It is by nature more predictive than is the case for existing plant where it is possible to carry out observation and measurement of human activity. On the other hand, HRA on new plant provides the opportunity to design operating procedures and standards which are adequate to meet reliability targets. Hopefully such studies may be timely enough to allow good human factors engineering to influence the design of the man/machine interface.

Classification of Fault Scenarios

Human Reliability Analysis is carried out on one or other of two plant operating conditions - normal or abnormal. It is these conditions which very largely determine the types of human error which are dominant in the analysis.

1. Normal conditions. The types of activity which are important during normal operation are:-

   - routine control
   - maintenance (preventive and corrective)
   - calibration and testing
   - restoration after maintenance
   - checking, inspection etc.

   Under these conditions there is normally very little decision making since most of the activities and the way in which they are carried out are determined by plant policy. Most of the errors can therefore be described as errors of commission (ECOM) or errors of omission (EOM).

   These errors all occur as discrete events within set sequences of activity.

   It is also notable that these types of errors may often be unrevealed at the time they occur and only become apparent as a result of a demand on equipment or a change in plant operating conditions. Errors of this type have been referred to as latent errors or 'pathogens' which are resident in a system and awaiting a local trigger to actuate them [3]. Because of the time lag between occurence and discovery, the error may be much more difficult to identify as the root cause of a problem requiring urgent remedial action. Root cause analysis is a science (or art) in its own right. The pursuit of root causes during accident or other abnormal plant operating modes may even lead to further successive errors as a result of misdiagnosis. Many serious accidents have arisen in this way and the potential run-on effects of latent errors arising from normal operation should be carefully considered.

2. Abnormal conditions. Activities under abnormal conditions fall into the following categories.

   - recognition and detection of fault condition
   - diagnosis and decision making
   - recovery actions

   Errors occuring during these activities tend to fall into three types:-

   a) Internal input. Errors of perception and discrimination

   b) Cognitive processing. Errors of interpretation, diagnosis and decision making

   c) Responses. Errors of omission and commission

   Often, under abnormal conditions there is the concurrent incidence of stress which will degrade human reliability.

Classification of Human Errors

The dominant types of human error present in a situation largely determine the HRA modelling technique which should be adopted. From the above, it is possible to see two dominant types emerge.

CLASS 1. Manual and sensory errors encompassing errors of omission, commission, selection, perception, discrimination etc.

CLASS 2. Cognitive errors arising from mental processing of information obtained from specific process indications combined with internal mental models, rules and learned responses e.g. diagnostic and decision making processes.

There are many taxonomies of human error which provide much more detailed methods of classification than this but which do not necessarily provide any guidance to the assessor in selecting a methodology. All classifications involve a degree of overlap and include fuzzy areas where it is difficult to fit human errors into one type or another. Such a grey area also exists in the two-fold classification described above. However, it has been found in the course of many HRA studies that this division provides a useful starting point for carrying out a human reliability analysis in a wide range of applications.

It has been found particularly appropriate to the study of abnormal operating situations as described above, where activities tend to fall into three phases, recognition - diagnosis - recovery. Recognition and recovery errors predominantely fall into CLASS 1 type errors. Diagnostic errors are essentially CLASS 2 type errors.

Modelling of Human Errors

CLASS 1 Errors are conveniently modelled using the methodology described by the THERP Handbook [1].

The origins of the present day THERP methodology lies in extensive work carried out by Sandia National Laboratories in the early 1950's in attempting to establish the quantitative influence of 1st order human errors on the effectiveness of military systems. The work was expanded in the 1960's to consider the human component in more detail. The work was later applied by the US Nuclear Regulatory Commission to the WASH 1400 Nuclear Reactor Safety studies [4]. Since WASH/1400, important refinements in method and data have been added. These include models for:-

1. Dependency between persons and within persons particularly during recovery actions.

2. Perception and response. An example of this is the Annunciator Response Model.

3. Stress and its effect on human reliability, particularly in abnormal or emergency situations.

In addition to these developments, THERP also includes methods for handling human error distributions by quoting upper and lower uncertainty bounds to allow for variability in human performance.

The purpose of THERP is stated to be two fold:-

- To present methods of modelling and estimating Human Error Probabilities to produce quantitative and qualitative assessments of human error in hazardous plants affecting availability or reliability of engineered safety features.

- To aid recognition and identification of error likely situations.

Steps in the THERP procedure are carried out interactively as follows:-

1. Define system failures (usually by means of a Probabilistic Risk Assessment)

2. List and analyse human interactions

3. Estimate probabilities of error using Human Reliability Analysis (HRA)

4. Incorporate HRA into PRA

5. Recommend changes and re-calculate the effects

THERP uses a special form of logic diagram (HRA Event Tree) to model human activities. The advantage of an Event Tree approach is that it is able to handle time-driven sequences of activity using conditional probabilities. An example of an HRA Event Tree is given in Figure 1. This example is based on the THERP approach but includes a full tabulation of the human errors identified in the sequence of activities. Care must be taken in drawing the tree to ensure that the errors identified are placed in a logical order in the sequence. For instance, it is not possible for the operator to misread an annunciator tile before he has detected it, or after he has failed to detect it. Most of these rules are based on simple logic or common sense, but in more complex sequences and error paths, it is possible to place errors in the wrong order. At each node of the tree, there is a chance of success or failure. The success path is represented by lower case and the failure path by upper case letters. In the simple HRA Event Tree in Fig. 1

FIG.1 HRA EVENT TREE - OPERATOR CHECK ON PLANT CONDITIONS

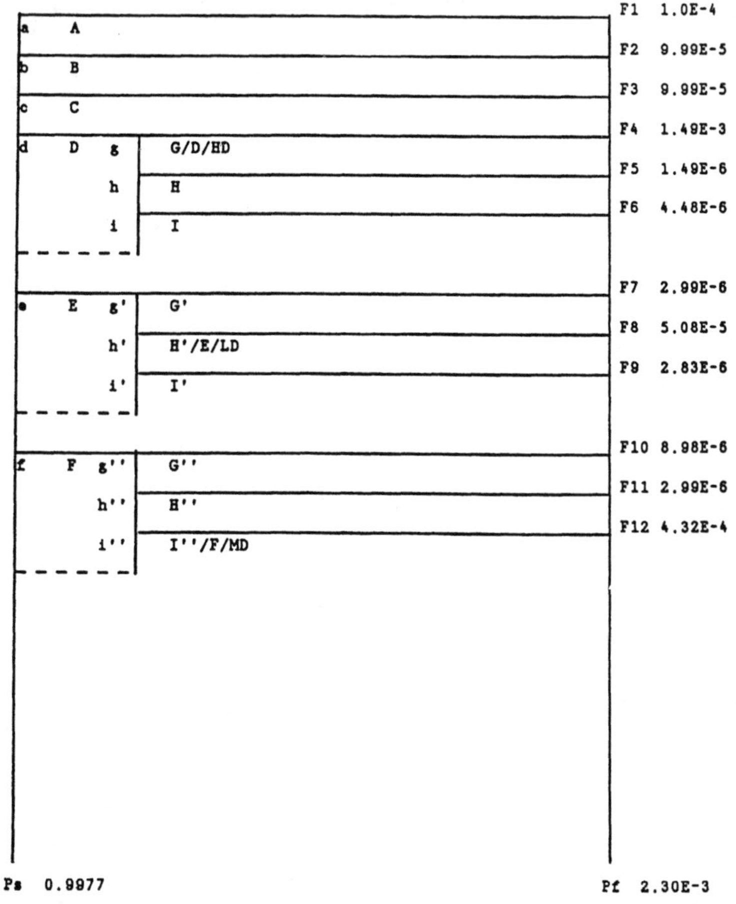

Left diagram labels:

a	A		F1 1.0E-4
b	B		F2 9.99E-5
c	C		F3 9.99E-5
d	D	g G/D/HD	F4 1.49E-3
		h H	F5 1.49E-6
		i I	F6 4.48E-6
e	E	g' G'	F7 2.99E-6
		h' H'/E/LD	F8 5.08E-5
		i' I'	F9 2.83E-6
f	F	g'' G''	F10 8.98E-6
		h'' H''	F11 2.99E-6
		i'' I''/F/MD	F12 4.32E-4

Ps 0.9977   Pf 2.30E-3

EVENT REF NO	ACTION		Ps / Pf
10	a / A	OPERATOR ACCEPTS BUT FAILS TO RECALL	0.9999 / 1.0E-4
11	b / B	OPERATOR MISREADS ANNUNCIATOR TILE	0.9999 / 1.0E-4
12	c / C	OPERATOR ASSUMED SPURIOUS (NO CHECK)	0.9999 / 1.0E-4
13	d / D	OPERATOR FAILS TO CHECK FAN PRESSURE	0.997 / 3.0E-3
14	e / E	OPERATOR SELECTS WRONG PRESSURE	0.999 / 1.0E-3
15	f / F	OPERATOR MISREADS FAN PRESSURE	0.997 / 3.0E-3
16	g / G	OPERATOR FAILS TO CHECK COLUMN D.P.	0.50 / 0.50
17	h / H	OPERATOR SELECTS WRONG D.P.	0.999 / 1.0E-3
18	i / I	OPERATOR MISREAD COLUMN D.P.	0.997 / 3.0E-3
19	g' g'' / G' G''	AS 16	0.997 / 3.0E-3
20	h' / H'	AS 17	0.949 / 5.1E-2
21	i' / I'	AS 18	0.997 / 3.0E-3
22	h'' / H''	AS 17	0.999 / 1.0E-3
23	i'' / I''	AS 18	0.855 / 1.45E-1

all the failure paths lead to a failure of the top event and are therefore additive. This may not always be the case, when some failure paths will be null paths - these should be shown but the Human Error Probability (HEP) need not be assessed.

The two most important features to be noted in the HRA Event Tree in Fig.1, concern recovery and dependency.

Recovery is shown from failure paths D, E and F via recovery success paths g, h and i. This is because the operator has a second chance of detecting a plant failure using a second diverse instrument reading. However at each node in the recovery path, there is the chance of failure G, H and I occurring and these are of course additive to the top event probability.

Whenever a recovery is possible as at node DG in Fig. 1, it should be considered whether any dependency exists. In this case there is estimated to be a high dependency (HD) between Action Dd (operator fails to check fan pressure) and Recovery Action Gg (operator fails to check Column D.p. (Differential Pressure). This is shown as G/D/HD, or having failed to Check Fan Pressure (D) there is a high probability he will fail to check column D.P. (G). THERP provides a dependency model to quantify this and other dependency levels.

The selection of Human Error Probabilities (HEP) shown in the Fig. 1 table is made by reference to the THERP Handbook Data Tables. These tables are conveniently summarised together with a search scheme in Chapter 20 of the handbook [1].

In spite of THERP being pre-emimently a Data Base method, matching of the human error to be quantified with an HEP value and description in the tables has a high degree of subjectivity associated with it. It is rare that the assessed task error and the Data Base error are identical. In any case the surrounding performance shaping factors (PSF's) such as level of training, stress, etc., will differ and need to be assessed. THERP provides a suggested Error Factor (EF) for each quoted HEP so that the range between upper and lower uncertainty bounds (UCB's) varies by a factor of (EF)^2. If the PSF's for the assessed task are considered relatively normal for that task and compare well with the Data Base PSF's, then the median figure can be used. By these means, THERP is capable of modelling and quantifying most of the Human Errors falling within CLASS 1.

CLASS 2 Errors of the Cognitive type may be modeled using the Human Cognitive Reliability (HCR) Model. This has recently been developed in the USA out of a research project sponsored by EPRI (Electric Power Research Institute) and undertaken by NUS Corporation [5]. It was developed to fulfil a need within the US Nuclear Power Plant (NPP) Industry to be able to quantify control room crew responses during complex accident sequences. The data used as a basis for the model was largely obtained from the results of NPP control room simulator trials. Out of this work an interim model has been developed [2]. The research project is to continue into 1989, and it is expected that further validation and statistical refinements will be made [5].

The basis approach adopted in the development of the HCR model, is that
during accident sequences, operating crews have a limited time within which
to make a correct diagnosis and perform key actions. The probability of
non-response within the allowed time period is described by a series of
correlations represented as time-reliability curves (Fig.2).

These curves have been associated with the three types of human cognitive
processing identified by Rasmussen [6]. These are known as skill-based,
rule-based and knowledge-based cognitive processing.

The other key inputs to the model are the median response time of the
operating crew, the time window available for diagnosis and the various
Performance Shaping Factors (PSF's) affecting the response.

Median response time (T) is the estimated time taken by the crew to complete
the actions following a stimulus. This time can be derived from simulator
trials or by observation in real plant operating situations. This median
response time is adjusted by coefficients for three important PSF's, each
performance shaping factor being modelled by a coefficient representing
quality.

The value of adjusted median response time is then given by T_A as in
equation (1) below.

The time window (t) for the diagnostic (cognitive) response of the operator
is a largely process determined parameter. Usually it is necessary to
calculate this time window by subtracting the sum of the time taken for all
other tasks from the total time available to the operators to safely
terminate the transient. This often takes the form of:-

    t = Process time available - (time to detect transient
          has occurred + time to terminate transient)

This assumes that the sequence of operations can be decomposed into separate
activities under the headings of Recognition-Diagnosis-Recovery as described
above.

The "normalised" time is given by the ratio t/T_A which is incorporated into
a mathematical correlation approximating to a three parameter Weibull
distribution.

$$P(t) = \exp\left\{-\left[\frac{t/T_A - \gamma}{\eta}\right]^{\beta}\right\} \tag{1}$$

where P(t) is the non-response probability.

In reliability engineering, the Weibull distribution has the advantage that
by adjusting the three shaping parameters (γ, η, β) it can be made to fit
many distributions of time to failure [7].

In the case of the HCR model, the shaping parameters are determined by the
predominant type of cognitive processing employed i.e. skill, rule or
knowledge based. At the present time, interim values of the shaping coefficients are available for use in the HCR correlation. These have been
derived from simulator trials and small scale tests [8].

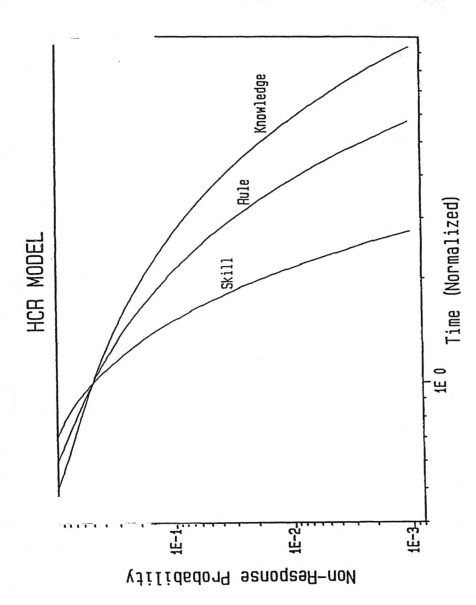

Figure 2 Normalized Crew Non-Response Curves for Skill-, Rule-, Knowledge-based Cognitive Processing

It should be noted that P(t), non-response probability, is not just the probability that no action is taken but that insufficient time is available to complete the task successfully, for various reasons e.g slow processing, inadequate information, wrong deduction etc.

The HCR model has been applied in a number of probabilistic safety assessments (PSA) studies for NPPs in the USA, none of which are presently reported. The technique also figured largely in a recent reliability benchmark exercise carried out under the auspices of the Commission of the European Communities Joint Research Centre [9]. The author has applied the method successfully in a process-industry application to diagnostic and decision making responses to process malfunctions where a severe time restriction is present. Although the results were found to be highly dependent on the response time and time window selected it is believed that the model provides answers which correlate well with engineering judgement when it is used sensibly in the type of application for which it is intended - human activities carried out under conditions of time restraint.

Continuous development of the model is still taking place, with the collaboration of six U.S. Electric Power Utilities, EPRI and Electricité de France (EdF) [8], [10]. This will include the development of additional scenarios and an exchange of data from simulators. A more detailed consideration of how the HCR correlation models the observed operator response, is also being developed. This Cognitive Sub Element (CSE) model will break down the HCR model into sub-events whose human error probability is in a range which can more easily be measured [10]. In this way it is fully expected that the HCR model will find an increasing role in human reliability assessment in the areas of cognitive reliability of diagnostic and decision making process.

COMBINED THERP/HCR MODEL

The combination of the THERP and HCR methodologies to give a unified approach can easily be achieved using the form of an Operator Action Tree as described in the SHARP Report [12]. An example of this is shown in Figure 3 in the form of a model which describes the main activities taking place in a holistic way. Clearly the OAT approximates more closely to the standard form of an Event Tree used for modelling accident sequences. The input to each event in the OAT is either an HRA Event Tree (as in Fig. 1) for CLASS 1 errors or the HCR model for CLASS 2 errors.

The contribution of each activity in the OAT to the top event (probability of failure of the complete sequence) can clearly be identified. In the example shown in Fig. 3, Event Reference No. 2 is the diagnostic response, the failure probability of which is derived from the HCR model. This is also the main contributor to the top event and clearly attention must be directed to the various parameters surrounding the cognitive response if the overall failure probability of the sequence is to be reduced. The example also shows that the re-checking of the plant conditions following the remedial action is a necessary part of the procedure but has little effect on the magnitude of the top event failure probability.

135

The way in which the main events comprising the sequence of activities are isolated for inclusion in the OAT is quite important. The first requirement is that the events should as far as possible be s-independent of one another in order to allow representation of recovery actions. An example of this is shown in Figure 3 where event 4 is a recovery action in the event of a failure in event 3.

CLASS 1 errors are shown such that each event is a holistic description of a series of sub-task errors decomposed in the form of an HRA Event Tree as in Fig. 1.

CLASS 2 Cognitive errors are generally isolated holistically and then quantified using the HCR model. Development of this model to include cognitive sub-elements [10] may well enable a form of decomposition of cognitive responses akin to that used by THERP. This will overcome the current lack of scrutability of the result which, although based on documented trials, provides no insight into the internal factors contributing to that result.

Where more than one cognitive response occurs within the sequence of activity, it is necessary to allocate portions of the available time-window to each response. Two methods are possible to resolve this problem. A subjective assessment may be made of the available time allocation as perceived by the operator given his set programme of activity.

Alternatively it is possible to use a convolution integral formulation where the success probability of the current task within the remaining time window is conditioned by the success probability in all the previous tasks. The result is an optimization of the overall success probability which effectively "mimics" the desired operator response [11].

The major advantage of the combined THERP/HCR model as described is its flexibility in modelling a very wide range of scenarios. The hierarchical decomposition of the human activity into major events using an Operator Action Tree enables structured procedures to be analysed for existing plants and developed for new plants. The further decomposition of main events into sub-tasks enables human error sequences to be identified. Quantification using either THERP or the HCR model as appropriate enables realistic human error probabilities to be assigned to all the nodes of the Event Trees. Optimum overall reliability for the total sequence of activities can then be achieved by modification of procedures and task design and by the application of ergonomic principles to the areas of need as indicated by the results of the analysis.

CONCLUSION

Current trends indicate that the inclusion of Human Reliability Analysis in PRA studies will become the rule rather than the exception. HRA practitioners are faced with a confusing choice of methodologies which may be used in assessments. The method chosen will to some extent depend on the resources and skills available as well as upon the nature of the scenario to be assessed. The level of decomposition which is possible in terms of information and time available will also have an influence. Ultimately,

FIG. 3 - OPERATOR ACTION TREE

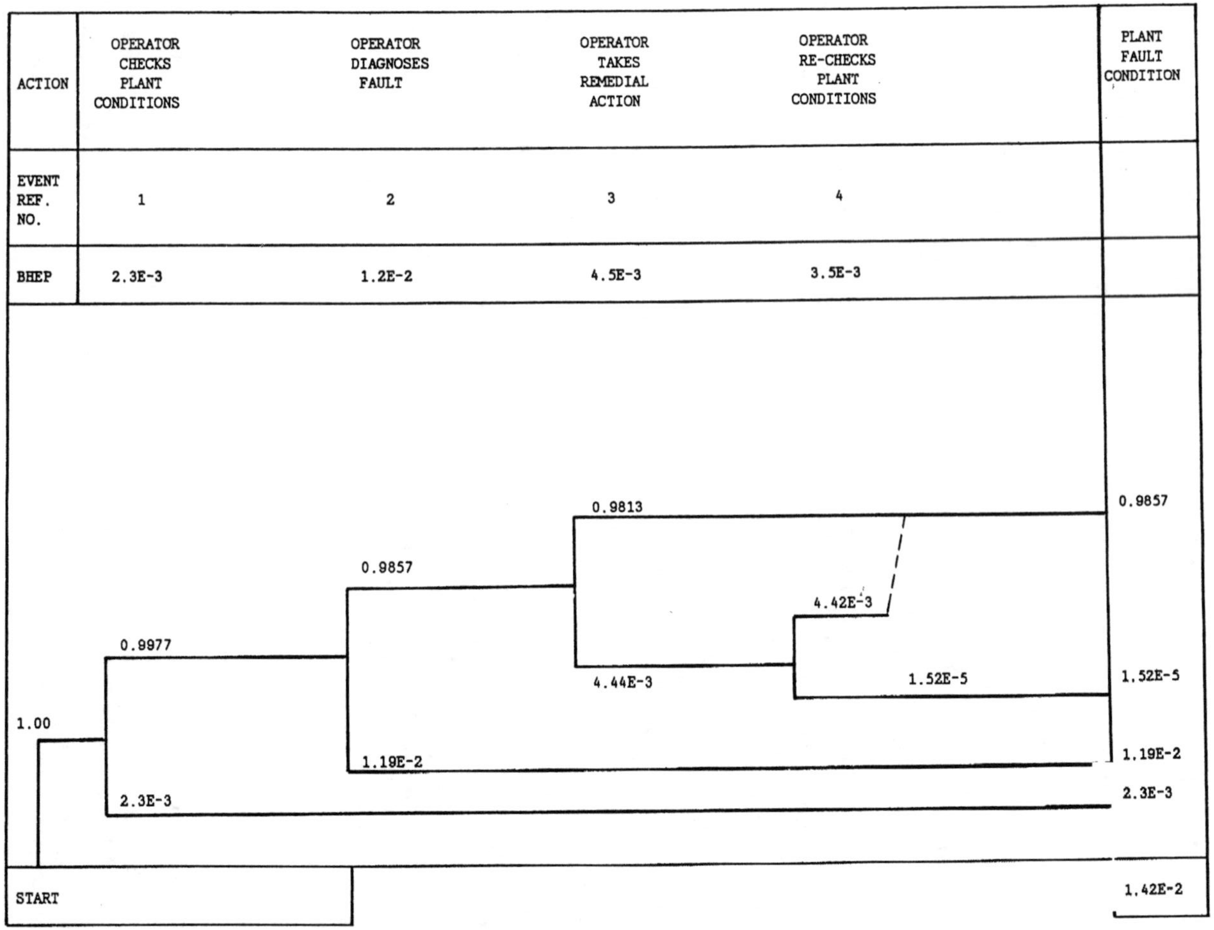

ACTION	OPERATOR CHECKS PLANT CONDITIONS	OPERATOR DIAGNOSES FAULT	OPERATOR TAKES REMEDIAL ACTION	OPERATOR RE-CHECKS PLANT CONDITIONS	PLANT FAULT CONDITION
EVENT REF. NO.	1	2	3	4	
BHEP	2.3E-3	1.2E-2	4.5E-3	3.5E-3	

0.9813

0.9857

0.9857

0.9977

4.42E-3

4.44E-3

1.52E-5

1.52E-5

1.00

1.19E-2

1.19E-2

2.3E-3

2.3E-3

START

1.42E-2

however, the choice of method must be dictated by such factors as credibility and scrutability if the results are to be fully defensible when subjected to critical appraisal. It is suggested that the approach described by this paper provides sufficient insight into human interactions to provide a defensible model of reality. By a series of decompositional steps, it is not only possible to model and quantify human error, but equally important, to indicate the deficiencies in human systems which make the dominant contribution to accident frequencies. By this means, attention can be given to areas of greatest need in terms of safety and cost effectiveness. The scrutability of the method ensures not only that human error has been fully accounted for but that this can also be fully demonstrated to regulatory authorities and others with an interest in plant safety.

REFERENCES

1. Swain, A.D. and Guttmann, H.E., *Handbook of Human Reliability Analysis with Emphasis on Nuclear Power Plant Applications*. Final Report. Sandia National Laboratories SAND80-0200. Prepared for US Nuclear Regulatory Commission. NUREG/CR-1278, 1983.

2. Hannaman, G.W., Spurgin, A.J., Lukic, Y.D. *A Model for Assessing Human Cognitive Reliability in PRA Studies*. NUS Corporation, presented at Third IEEE Conference on HUman Reliability, Monterey, California, 1985.

3. Reason, J.T. *The Human Contribution to Nuclear Power Plant Emergencies*. Conference proceedings. Human Reliability in Nuclear Power. London, 1987.

4. NUREG-75/014. WASH1400. *Reactor Safety Study - An Assessment of Accident Risk in US Commercial Nuclear Power Plants.* US Nuclear Regulatory Commission, Washington, DC. 1975.

5. Joksimovich, V., Worledge, D.H. *Using Simulator Experiments to Analyse Human Reliability for PRA Studies*. Article: Nuclear Engineering International. January 1988, p37-39.

6. Rasmussen, J. On the structure of knowledge. *A Morphology of Mental Models in a Man-Machine Context*. RISO-M-2192, Roskilde, Denmark. 1979.

7. O'Connor, P.D.T. *Practical Reliability Engineering*. John Wiley and Sons. Second Edition 1986. P39-40.

8. Hannaman, G.W., Worledge, D.H. *Some Developments in Human Reliability Approaches and Tools*. International Post - SMiRT 9. Seminar on Accident Sequence Modelling, Munich, FDR, August 1987.

9. Commission of the European Communities. *Report of the Human Reliliability Benchmark Exercise* held at JRC, Ispra, Italy. To be published in the Autumn of 1988.

10. Lukic, Y.D, Worledge, D.H., Hannaman, G.W., Spurgin, A.J. *Modelling Framework for Crew Decisions During Accident Sequences.* International Conference on Advances in Human Factors in Nuclear Plant Systems, Knoxville, Tenessee, 1986.

11. Vestrucci, P. *A Convolution Approach to Assessing Crew Response Probability for a Sequence of Actions with an Overall Time Restraint*. Private communication to C.E.C. HR - Benchmark Exercise participants.

12. Hannaman, G.W., Spurgin, A.J. *Systematic Human Action Reliability Procedure (SHARP)*. NUS Corporation. Prepared for EPRI. Research Project 2170-3. Report No. NP-3583. June 1984. Page 3-33, 34.

139

A PRACTICAL APPLICATION OF QUANTIFIED RISK ANALYSIS

G Purdy
Technology Division
Health and Safety Executive
Bootle
Merseyside, UK

ABSTRACT

Quantified Risk Analysis (QRA) techniques have been developed in HSE to assist decisions, on the desirability of developments near hazardous installations based on numerical estimations of Individual and Societal risk. These techniques can also be used to measure the effectiveness, in terms of relative risk reduction, of changes to plant and procedures, including methods of work and the emergency response of workers.

A case study demonstrates how simple event –and fault– trees can be used to synthesise event frequencies comprising both hard-ware faults and human error. The 'robustness' of the trees is established by sensitivity analysis. The analysis is used as a means to suggest improvements in plant standards and working methods; the performance criteria for these and the overall reduction in risk levels to members of the public nearby once the changes have taken place, can be estimated.

INTRODUCTION

Quantified Risk Analysis (QRA) techniques have been developed in HSE to assist in reaching informed decisions on the desirability of developments on or in the vicinity of Major Hazard plants, based on numerical estimations of Societal and Individual Risk. The methodology is related to techniques developed in the Nuclear Industry over the last 25 years. In that industry, QRA has been used not only for planning and conceptual studies but to optimise the level of safety of existing plants by identifying 'cost effective' remedial measures (1). There have, however, been few public presentations of the use of QRA techniques to improve the safety of Major Hazard plants (2).

With nuclear plant, there can be a great deal of similarity between the installations and the type of risk they pose to the surrounding population. This allows the results of analyses for different plants to be compared and makes the job of setting standards easier. With chemical plant much more care is needed to maintain a consistent approach for different installations and there are great difficulties in comparing risk results from different studies. HSE has recently published a discussion document on the tolerability of risk from nuclear power stations (3); a subsequent document will discuss risk criteria for use in land-use planning around major chemical hazards.

QRA can in principle be used in an absolute mode (ie, to show the overall risk level from a plant) or in a relative mode (ie, to compare options) to understand what are the major contributors to the overall risk from a major hazard. It can also be used to show the benefits, in terms of risk reduction, that accrue from changes in plant and working methods. Obviously the methodology and assumptions must be consistent for such 'before and after' calculations.

The Health and Safety at Work etc Act 1974 (4) and associated legislation place duties upon an employer that are qualified by the phrase, 'so far as is reasonably practicable'. The onus is placed on the employer to prove that a measure is not reasonably practicable; to do that requires a consideration of the level of risk and the costs to offset it. Only when the risk is insignificant in relation to the cost does the employer discharge his duty (5). Normally, decisions on what is reasonably practicable are based on codes of practice, standards and good industrial practice. Where these are not available a qualitative judgement is needed, based on professional expertise augmented with whatever quantification is possible. Using QRA it is possible in some cases to estimate directly the reduction in risk brought about by a specified expenditure. Whilst the comparison of reductions in risk to financial costs remains controversial, a statement that, for instance, the level of individual risk to residents at 500 m can be reduced by 2 orders of magnitude for an expenditure of £10,000 may well help in deciding whether improvements are reasonably practicable. This paper uses a case study to demonstrate this approach.

One of the most difficult tasks facing the risk analyst is to obtain appropriate failure rates, especially for human actions. There have been significant advances made in the assessment of human factors, especially in quantifying errors in skill-based behaviour by methods which seek to produce answers that are independent of the assessors experience or the method used. It is not easy, however, to build human error and plant failures into the same fault tree; great care is needed to deal with commonality. We have been able to speed up this process by using a multi-disciplinary "committee" to decide on upper bound, best estimate and lower bound base event frequencies. Only in those cases where sensitivity analysis showed great effect on the calculated level of risk, was a more rigorous approach applied. The overall robustness of the model was also tested by using all the 'worst case' and then the 'best case' event frequencies to determine the change in the calculated level of risk.

BACKGROUND TO THE STUDY

As the result of a request from a planning authority for advice on a draft local plan, HSE carried out an assessment of a liquefied toxic gas (LTG) installation. We used a limited case, 'top-down' HAZAN technique to produce a representative list of failure cases as an input to the HSE RISKAT risk analysis tool (6). Each failure case consists of a release rate (continuous) or size (instantaneous) together with a duration and probability. The intention is to select a range of discrete failure cases which characterise the risk from the installation. HSE has developed rules and assumptions to enable this to be carried out systematically and consistently for different installations. To this end, single values or in some cases ranges of values for failure and event durations are available; the assessor can justify any deviation from these because of specific plant conditions or the demands of the analysis.

The major purpose of the installation is to export LTG by tanker; a large number of these are filled each year. The simplified 'P&I' diagram for the loading operation is at figure 1. Other site specific features are the storage of LTG in large vessels at temperatures below ambient provided with walled spillage retention, short liquid lines with low flow rates and the provision of remotely operable shut-off valves (ROSOVS)' on important lines. It is likely that if a vessel or pipe containing LTG failed, the resulting release would be of low energy with very little flashing; the resulting evaporation rate from the contained pool would be low. Furthermore, as some pipework is only 'live' for a fraction of the year, the corresponding event frequencies would be low. This qualitative analysis suggests that the loading operation would make the highest contribution to offsite risk.

An operative, who works on or around the loading gantry, controls the loading operation (Figure 2). It is his job to vent down, couple up and fill tankers to a specified load. The tankers are filled through flexible hoses, these are held on a stanchion and released by using a captive key system interlocked with the closure of a barrier in front of the tanker and cab. The tanker driver goes to a rest room away from the tanker during loading. The flow of LTG can be stopped by the operative pressing a button on the wall of his cabin which closes the ROSOV up-stream. This ROSOV and other valves further up-stream, can also be closed from the control room but this is 100 m away and out of sight of the tanker filling bay. In the event of a significant release of LTG, the loading bay operative is faced with two alternatives:-

(i) if he is in the cabin he can attempt to close the ROSOV; or

(ii) he can attempt to run to the control room.

A breathing apparatus (BA) set for emergency use is located on the gantry outside the cabin.

THE ANALYSIS

Standard practice is to carry out an initial QRA to find the dominant failure cases and their contribution to the risk. These are then subject to more critical study to improve the accuracy of the final result (sensitivity analysis). The initial analysis, in this case, showed that the calculated levels of Individual Risk (IR) were high and that releases due to potential events at the tanker loading bays dominated the off-site risk; at 500 m these were 95% of the total.

A critical factor was the possible duration of any release. We considered three possible durations for such an event:

1 minute representing 'immediate' operation of the ROSOV;

5 minutes representing delayed operation of the ROSOV or other valve; and

20 minutes representing eventual closure of valves manually by an emergency team.

Gas penetration calculations showed that in the event of hose severance, the concentration of gas inside the cabin, even if the door was shut, would exceed 1000 ppm within 10 seconds. At this concentration the operative would be instantly incapacitated and likely to die. Given the speed at which the cloud would travel, the operative was also likely to die if he was caught outside unprotected. Nevertheless there remained a small chance that the duration of the release could be limited by closure of the ROSOV from the control room or by the very rapid response of the operative. The basic event tree (Figure 3) was drawn up to proportion the overall event frequency between the 1, 5 and 20 minute releases. The only mechanism whereby the release could be stopped within 1 minute would be for the operative to be in his cabin and to react instantly. The release could be stopped in 5 minutes if:-

a) the operative successfully ran to the control room and either he or the staff there successfully closed the ROSOV;

b) a second operative on duty outside noticed the release, survives, ran to the control room and warned the staff there, who in turn operate the ROSOV;

c) the operative recovered from his initial inclination to panic, doned BA and closed the ROSOV.

In other cases the release is assumed to be of 20 minutes duration.

From Figure 3 the probabilities of each duration of release are:-

$$P_1 = P_f \cdot N \cdot (1 - P_a) \qquad\qquad (1)$$

$$P_5 = P_f \cdot N \cdot P_a \cdot (1 - P_b) \qquad\qquad (2)$$

$$\& \quad P_{20} = (P_f \cdot N) - (P_1 + P_5) \qquad\qquad (3)$$

Where

P_f is the coupling failure rate per loading, and

N is the number of loadings per annum.

Figures 4 and 5 are fault trees drawn to represent the paths to successful closure of the ROSOV within 1 or 5 minutes respectively. Implicit in the trees are the following assumptions:-

a) that if the operative has to don BA, this takes more than 1 minute;

b) if the operative is out of doors at ground level, he will not/cannot reach the operating button in his cabin before he is overcome; and

c)   the second operative can only stop the release by going to or contacting the control room; this takes more than 1 minute.

A full Boolean reduction was applied before quantification to produce minimum cut sets and to deal with commonality. The trees were drawn and values for the 21 base event probabilities were arrived at by discussion and agreement within a committee comprising members with a range of experience in human factors, risk assessment and chemical plant standards. Wherever possible, hardware failure and human reliability data were obtained directly from or by applying judgement to recognised data banks and sources of data in other studies (7)(8). For each base event, upper (UB) and lower bound (LB) and best estimate (BE) values were chosen; these are given in Table 1.

To facilitate the sensitivity analysis, computer program was written which allowed the user to input base event probabilities. The top gate probabilities were then fed into the event tree (Figure 3) which provided three release cases for input to RISKAT. These together with those from other events were used to derive the overall level of IR from the installation. Figures 6 and 7 show the results using the worst case, best case and best estimate values for base event probabilities. The closeness of the curves and of the risk contours are felt to demonstrate the robustness of the method. The overall risk level is stable and unaffected by reasonably perturbation of the base event probabilities.

However, by selecting values outside the UB and LB, the effect of significant changes in plant or working methods could be demonstrated.

It was found that the results were sensitive to the proportion of the time the operative stayed in his cabin, his tendency to panic or the

TABLE 1

Base Event Probabilities

Event No.	Description	Justification	BE	UB	LB
1.	Operator outside. Injured either immediately or on way to Cabin or CR within 1 minute.	High prob. of injury from guillotine failure. Closeness to leak. Canister respirator only.	0.9	1.0	0.8
2.	Operator inside. Injured (so does not press button within 1 minute.	More protection than out of doors. Orthogonal jet orientation.	0.2	0.5	0.1
3.	Operator outside. Panics and does not reach CR within 1 minute.	Pure panic – freezes. Judgement.	0.5	0.9	0.3
4.	Operator inside. Panics so does not press button within 1 minute.	½ outside figures for BE.	0.25	0.9	0.01
5.	Operator gets to CR. They fail to respond correctly within 1 minute.	100 m to CR. Operator may be injured/confused. Assumes high level of training.	0.01	0.1	0.005
6.	Operator in cabin.	15 minutes in 1 hour estimated out of doors.	0.75 (Best Case)	0.9	0.5 (Worst Case)
7.	Local isolation system works.	Lees (7) – mech. failure rate of valve on demand used.	0.999	–	–

TABLE 1 (cont)

Base Event Probabilities

Event No.	Description	Justification	BE	UB	LB
8.	Guillotine failure occurs.	3×10^{-6} per operation F. rate used.	9.885×10^{-3}	–	–
9.	Isolation systems operated for CR fail.	As 7 above.	0.999		
10.	Operator in cabin. Puts BA on before pressing button. ie 1 minute.	Passes button to get BA. But panics.	0.1	0.5	0.01
12.	Operator is outside. Injured within 5 minutes and doesn't reach CR.	cf 1	0.9	1.0	0.8
13.	Operator is outside, panics (not injured) and does not reach CR at all.	$^1/_{10}$ of 3	0.05	0.1	0.03
14.	2nd Operator fails to detect leak.	Judgement	0.5	0.75	0.1
15.	2nd Operator detects a leak but is injured and CR not contacted.	NB: Tannoy System available.	0.01	0.1	0.002
16.	2nd Operator detects leak but panics so CR not told.	Judgement	0.01	0.1	0.001
18.	Operator is inside. Not injured, panics and remains uninjured. Recovers; does he put BA on?	High success having survived and recovered.	0.9 (Best Case)	1.0	0.5 (Worst Case)

148

TABLE 1 (cont)

Base Event Probabilities

Event No.	Description	Justification	BE	UB	LB
19.	BA fails	10^{-4} demand valve failure.	10^{-3}	–	–
20.	Operator panics when attempted isolation fails.	High prob when sees failure.	0.8	1.0	0.1
21.	Operator is inside and panics for 5 minutes	½ of 4	0.02	0.1	0.01

likelihood that he will be incapacitated by the LTG. The level of risk from this site was largely dependant on the actions and location of this operative who, in the event of a release of LTG, would be expected to act in a way that reduced his own chance of survival. Any improvements to the plant could therefore sensibly concentrate on either protecting him more to optimise his chance of closing the valve, building in redundancy (ie, having two men on the job) or preferably by providing an automatic detection and valve closing system. The model allowed us to explore the benefits of a range of theoretical changes to the plant. The results are summarised in Figure 8; the cases chosen for comparison with the base case are:-

> a. Auto – this represents a device that on sensing a leak initiates, without human intervention, the ROSOV with a system failure rate of 10^{-3} per demand (this is not considered unreasonable for a well engineered system).

b. Alarm in Control Room – a detector causes an alarm to ring in the constantly manned control room. The two cases show the difference if the operative there acts immediately to close the ROSOV or delays in so doing. The difference between the consequential risk levels calculated is large. This demonstrates the need for effective emergency training in which alarm conditions are simulated. A high level of skill on the part of the control room staff is essential to make the correct diagnosis and act promptly. One difficulty arises because of the tendency of gas detector systems to a high level of spurious alarms, which in turn lead to reduced operator response. However, we could show that while good Control Room operator response was important, detector reliability was less so; variation of the failure rate on demand between 10^{-2} and 10^{-3} caused little change in the overall rise.

c. Protecting the Operative – Three different strategies were tested: making his cabin a "gas tight" room, making him wear BA at all times and providing a second operative in a "gas tight" room. The latter was found to be the most effective option.

It would have been possible to test many other safeguarding strategies, but our intention was not to produce complete solutions but rather to demonstrate that a significant reduction in the level of offsite risk could be achieved by realistic changes to plant and to working practices. The analysis also showed how, for some strategies, operator response remained critical.

APPLICATION OF THE ANALYSIS

This work allowed a very constructive discussion with the occupier of the installation about modifications that could reduce levels of offsite risk. They responded with a package of measures that included:-

a. many more operating buttons for the ROSOV;

b. a gas tight cabin with a monitored air supply;

c. a 'pressure drop' monitoring system to sense leaks in the liquid gas line and filling hose which in turn would close the ROSOV and operate an alarm in the control room; and

d. either a gas detection and ROSOV closing system with voting (one for alarm, two to close valve) or a 'rate of increase in load' detector with voting.

The model was used to test the occupier's suggestions. The results are shown in Figures 9 and 10. The reduction in offsite risk levels is dramatic (over an order of magnitude at most distances). The contour plot (Figure 10) shows the very large reduction in the areas of land and therefore possibly the number of people affected.

We were also able to use the model to judge performance criteria for the various safeguarding measures. We found that:-

a. The operative no longer had a critical intervention role. The level of offsite risk was no longer dependant on whether he closed the ROSOV. Whilst he may stop minor leaks which the detectors might miss, his primary role would be to save himself. Such a response was felt to be more realistic.

b. The reliability of the ROSOV was not critical within realistic reliability boundaries; very little difference was found by varying the failure rate from 1 in 1000 to 1 in 100 per demand.

c. The reliabilility of the detectors was more important but not critical given well designed systems. Providing one of the systems was fairly reliable (10^{-2} demand^{-1}), there was little change when the detection rate of the other varied from 0.9 to 0.5 demand^{-1}. As the pressure drop system would measure the flow conditions directly this was expected to have a high reliability; the reliability of the gas detector system was therefore less important.

ACKNOWLEDGEMENT

I would like to thank my colleagues, Geoff Grint and John Brazendale, for their invaluable help in this work.

REFERENCES

(1) Wash-1400, Reactor Safety Study - An Assessment of Accident Risks in US Commercial Nuclear Power Plants, US Nuclear Regulatory Commission, Washington, NUREG-75/014.

(2) Guymer, P., Kaiser, G. D. and Mckelvey, Probabilities Risk Assessment in the CPI, Chemical Engineering Progress, January 1987, pp 37-45.

(3) Health and Safety Executive, The Tolerability of Risk From Nuclear Power Stations, HMSO, London, 1988.

(4) Health and Safety at Work etc Act 1974, HMSO, London, 1975.

(5) Asquith, L. J., in Edwards v National Coal Board, (1949) IKB 704.

(6) Pape, R. P. and Nussey, C., A Basic Approach for the Analysis of Risks for Major Toxic Risks, The Assessment and Control of Major Hazards, I Chem E Symp., 22-24 April 1985.

(7) Lees, F. P., A Review of Instrument Failure Data, I Chem E Symp. Series No. 47.

(8) Swain, A. D., and Guttmann, H. E., A Handbook of Human Reliability Analyses with Emphasis on Nuclear Power Plant Applications, NUREG/CR-1298, Washington DC 20555.

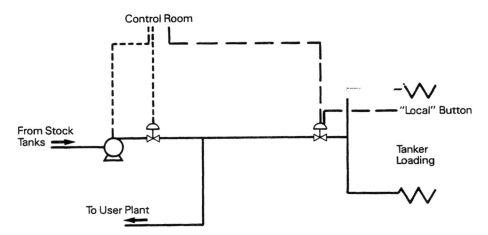

Figure 1 Simplified P & I Diagram

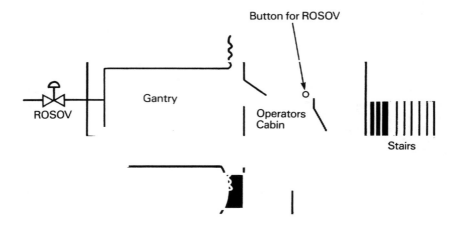

Figure 2 Tanker filling bay layout

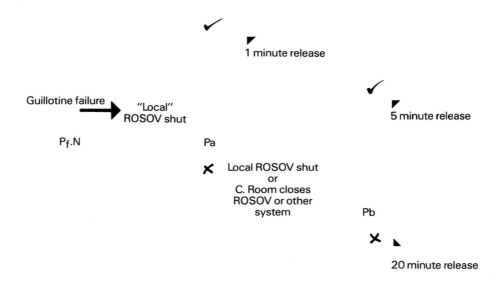

Figure 3 Event tree for isolation of tanker loading hose

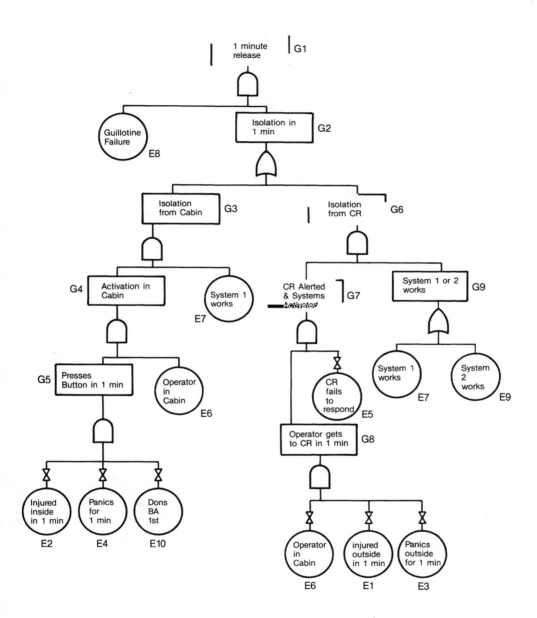

Figure 4 Fault tree for 1 minute release

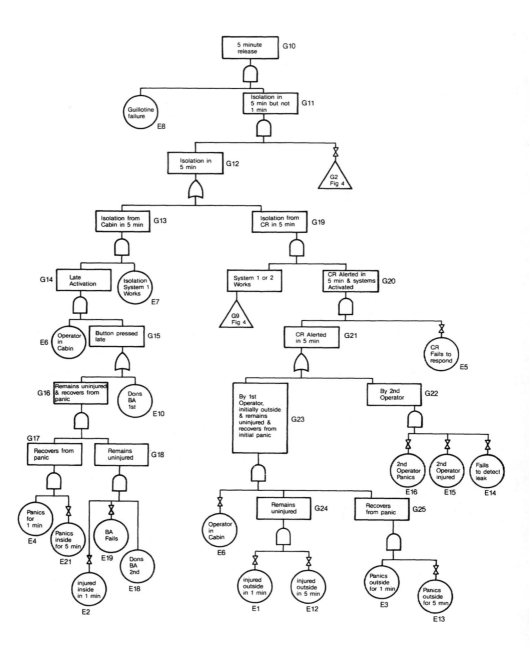

Figure 5 Fault tree for 5 minute release

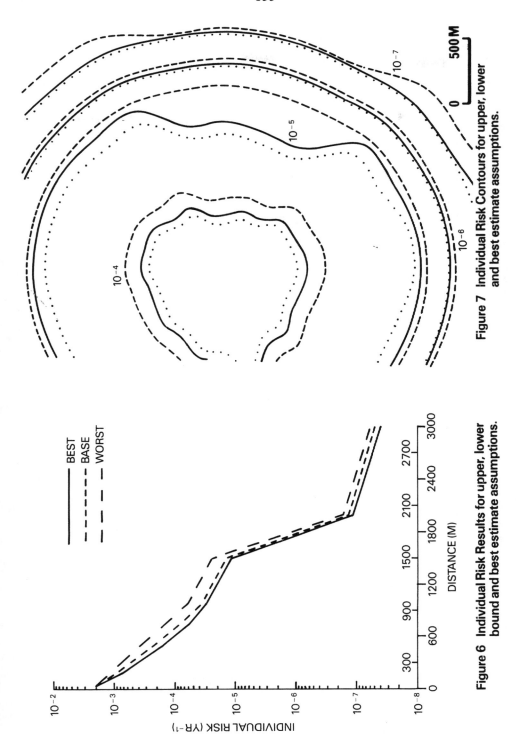

Figure 7 Individual Risk Contours for upper, lower and best estimate assumptions.

Figure 6 Individual Risk Results for upper, lower bound and best estimate assumptions.

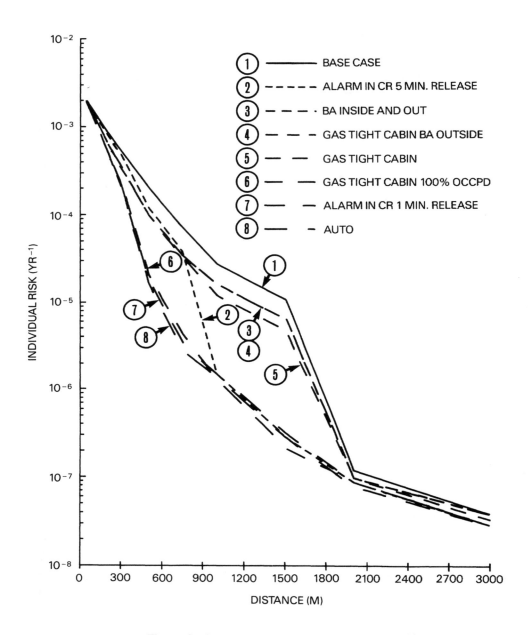

Figure 8 Comparison of installation individual risk after hypothetical improvements

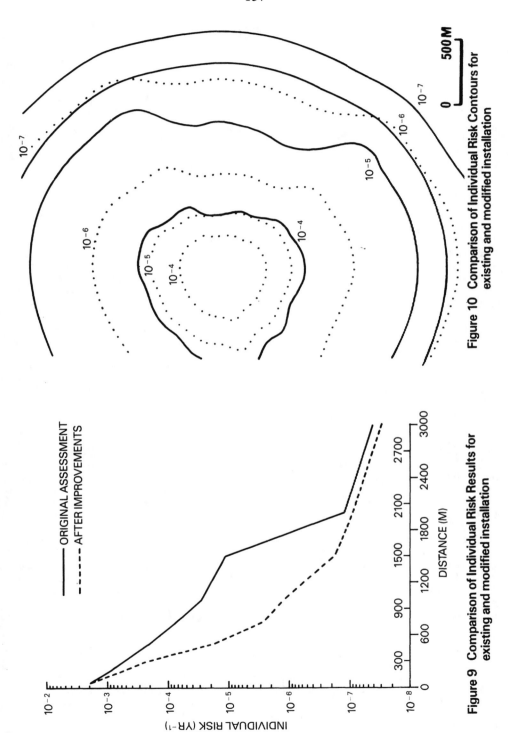

Figure 10 Comparison of Individual Risk Contours for existing and modified installation

Figure 9 Comparison of Individual Risk Results for existing and modified installation

PROF A COMPUTER CODE FOR PREDICTION OF OPERATOR FAILURE RATE

K.HARALD DRAGER & HELGE S. SOMA
A/S QUASAR CONSULTANTS
ODD FALMYR, STATOIL

ABSTRACT

PROF (PRediction of Operator Failure rate) is a computer code developed to calculate the operator reliability for different types of control room tasks. The calculated failure rate may be applied in a risk analysis or for the purpose of evaluating alternative designs for a man-machine system.

The program is developed as a tool to prevent human error, by identifying the factors that contribute to such errors. It is based on a set of operations, for which it has been possible to collect reliable data. These data come from either empirical studies or expert judgements, and make up the data base of the system. The program is designed so that updating of the data base, when new data are available, can be performed quite easily.

Based on this data base, a regression analysis is performed to predict the failure rate of the operator. To perform these calculations the program user needs to specify the values of a set of Performance Shaping Factors (PSFs), which involve contribution to such errors. These values may be given by experts based on the knowledge of the situation in which the task is to be performed.

The database is mainly made up in cooperation with Norsk Hydro and STATOIL, based on experiments performed at their simulators for Oseberg and Gullfaks control rooms, and reported accidents where it has been possible to estimate the PSF's for the situation in question.

Experience from use of PROF is presented.

INTRODUCTION

In performing a risk analysis one is often faced with the limited availability of data for the quantification of the system reliability. This is especially true when it comes to assessing the contribution of the operator to the overall system reliability.

In this paper an EDP code for PRediction of Operator Failure rates (PROF), which has been developed through a joint research project involving QUASAR and Norsk Hydro, is presented. The project has financially support from NTNF (Norwegian Council for Scientific and Technical Research), and the EDP code is at present beeing further developed in a joint project between Norsk Hydro, STATOIL and QUASAR.

The program may be used in a risk analysis, as an aid for assessing the contribution of the operator to a system failure, or as an aid for evaluating different design options for use in a control room.

For this second type of use, the program may be used as an aid for preventing design induced errors.

THE MODEL

The calculations of the operator failure rates are based on a regression analysis model. The operator failure rate may be stated as

$$\log f = \sum_{k=1}^{m} PSF_k \times W_k + C$$

or

$$\log f = PSF_1 \ W_1 + PSF_2 \ W_2 + \ldots \ldots \ldots + PSF_m \ W_m + C$$

where

f = the operator failure rate

C = numerical constant

PSF_k = numerical value of PSF_k

W_k = weight of PSF_k

m = number of PSF's

The PROF model contains 9 PSF's, i.e. $m=9$.

The calculations may be performed for a set of operations for which it has been possible to collect data from empirical studies or other types of accepted sources.

The data base contains data for the following operations.

```
*    Check reading and quantitative reading of
     displays
*    Reading of labels
*    Inspection on a change of shifts or plant
     walk through
*    Selection of manual controls
*    Actuation of manual controls - commision
*    Responding to annunciated legend lights
*    Completing a series of multiple tasks
*    Maintenance and calibration tasks
*    Communication tasks
*    Maintenance and calibration tasks
```

At present the data base is small, a fact which reflects the limited availability of data. However, in the project mentioned, the aim is to extend the database with data from simulations at the Gullfaks and Oseberg Simulators.

The data base contains data with different levels of subjectivity. To reflect the difference in data reliability the data has been weighted accordingly :

Weight	Type of data
0.8 - 1.0	Good experimental data
0.5 - 0.8	Empirical and subjective data
0.1 - 0.5	Subjective data

The category "Good experimental data" contains data from experiments where it has been possible to check out how the experiments were done, eventual weakness in methods etc. One such data set is data from the "Wellsim-experiments" performed at Rogalandsforskning, and from the Gullfaks and Oseberg Simulators.

Category number two, "Empirical and subjective data" will contain data from Swain & Guttman's "Handbook"/1/ and other such reports/2/.

The last category consists of data from expert judgements and other types of subjective data.

Experimental data seldom reflects more than one independent variable's (PSF's) influence on the operator performance. Without putting systematic errors into the data one must then handle the possible influence of other variables as well. In the model this has been taken care of in the following ways:

1. Non-specified variables are given medium values.
2. Non-specified variables are estimated by experts.
3. Non-specified variables are treated as unknown, which have to been given a value by use of a relevant method.

PERFORMANCE SHAPING FACTORS

The calculations are based on values given to the factors known to influence the operator's performance, the Performance Shaping Factors (PSF). The model does not cover all the PSF's presented in the literature /1/, but a number of the most important ones have been selected. The selection of PSF's was a result of the research project mentioned. The main objective in the selection process was to choose a number and types of PSFs possible to handle in a practical way. The choice must not be looked upon as a final selection, but as a "best choice" selection, given todays state of the art. These PSF's cover such factors as the operation, the man-machine interface, the situation and the operator.

The selected PSF's are

```
*    Task complexity
*    Information quality
*    Feedback
*    Breaking of stereotypes
*    Distractions
*    Time pressure
*    Exposure to danger
*    Task relevant experience
*    Personnel qualifications
```

Giving values to the PSF's is the main task of users of PROF. Normally this will be done in a group containing people with "expert" knowledge of the system to be evaluated. If no other scale is given, the values of the PSFs are set according to the following scale:

PSF subjective scale	PSF value scale
Optimal, much better than normal	1
Better than normal	2
Normal	3
Worse than normal	5
Far worse than normal	10

To give a PSF the value 1 or 10 may only be done in those cases where it is certain that it can hardly be better or worse.

PSF 1 – Complexity of the operation

For simple operations the above mentioned scale will be used. For more complex operations, however, the value is set by applying the following :

1. Error tolerance of the system (Et)
2. Cognitive workload (Cw), i.e. number of tasks, how much information to be remebered by the operator etc.

The PSF value will then be $\{1 + Et \times (Cw - 1)\}$

The error tolerance of the system is decided accordingly :

Error tolerance	Et value
Large margins, limited requirements of accuracy	0
Normal accuracy required	0.5
Limited margins, great accuracy required	1

and the cognitive workload :

COGNITIVE WORKLOAD	NUMBER OF TASKS			
	1 - 2	3 - 5	5 - 10	> 10
Little or no information processing required	1	2	3	5
Some inform. processing required	2	3	5	7
Considerable information processing required	3	5	7	10

PSF 2 – Information quality

The information quality term referes to the quality of the machine displays, regarding how unambigous the information from the machine to the operator is, when the operator is checking the system status. This PSF is given values according to the standard value set.

PSF 3 - Feedback

This PSF also concerns the quality of the man-machine interface, although in a different way to PSF 2. Feedback concerns the information the operator gets when operating the system, i.e. when the operator is operating the machine controls. The quality of the feedback depends on how unambiguous the information is and if there is a time delay involved.

Excellent feedback is what you get when, upon turning a switch, you get visual, tactile as well as auditive feedback to tell you that you have put the switch in the right position. If such a complete set of feedback is given the value 1 may be given. Visual feedback alone gives a value of at least 3, whilst lack of any of the other two sets the minimum value to 2.

If there is a time delay between the performance of the operation and the feedback, this may increase the PSF-value. This is especially true if more tasks are to be performed.

PSF 4 - Breaking of stereotypes

Breaking a populational stereotype or a standard way of presenting information to the operator may be done in several ways. (Using the color green to present an alarm, counterclockwise turning of a switch to increase the volume etc.). The values are given as

Breaking of stereotype	PSF-value
Complete accordance with standard	1
Minor deviation from standard	3
Severe deviation from established standard	5
Most serious deviation from standard	10

PSF 5 - Distractions

Noise, presence of other people, or other operations which are disturbing the operator whilst performing the actual operation make up PSF 5, which is given values according to the standard value scale.

PSF 6 - Time pressure

Time pressure as used in this model covers not only what is normally considered as time pressure, shortage of time, but also those cases where there is too much time available.

When assessing time pressure (T) one may consider it as the relation between the time available (Ta) and the time needed to perform the operation (Tp), Ta/Tp = T. If this method is chosen, the time pressure will get the following values

Time pressure	PSF value
T >=10	5
10 > T > 5	3
5 > T > 3	1
3 > T > 2	3
2 > T > 1.5	5
T <= 1.5	10

PSF 7 - Exposure to danger

This PSF reflects the danger the operators experience that they are exposed to. As such the value of this PSF should reflect the subjectively experienced dangers involved, and not the objectively calculated ones. Such danger may be direct, as an actual threat of death or injury, or indirect, as a production stop, getting fined, losing a job etc. Direct exposure will always be given higher values than indirect exposure. Experience of no exposure to danger may also be regarded as negative as this may lead to carelessness.

Exposure to danger	PSF value
No exposure to danger	3
Minor exposure to danger	1
Personal exposure to danger	5
Critical exposure for oneself & others	10

PSF 8 - Task relevant experience

The task relevant experience of the operator is defined as by how often the operator perform the task or a similar one. If sufficient information is available, one should also consider the time passed since the operator performed the task last. Values are given accordingly :

Interval between performances	PSF value
Often - daily	1
Regularly - weekly	3
Seldom - monthly	5
Rarely - yearly	10

PSF 9 - Personnel qualifications

Personnel qualifications cover a large amount of personnel characteristics, such as knowledge and insight to the process, organizational climate, formal education etc.

If the task to be evaluated is one in which regular procedures are to be deviated from, process-insight and knowledge should have a large impact on the PSF value. It is also advised that too low or too high formal education with regard to the job requirements should give a PSF value greater than 3.

Interdependence between PSF's

In order to provide correct results, the regression analysis variables should be independent of each other. The PSF "Breaking of stereotypes" could depend on important factors with respect to other PSF's like distractions, time pressure and exposure to danger. If the mentioned factors are included in more than one PSF when assigning values to the other PSF's, double counting should be avoided.

Insufficient knowledge of a PSF

One problem that may occur when using the method is that there may not be sufficient information available to give values to all the PSF's. In the PROF model one has chosen to solve this problem by giving the PSF a value that will not have significant influence on the overall prediction.

THE PROGRAM SYSTEM

The system has the following layout

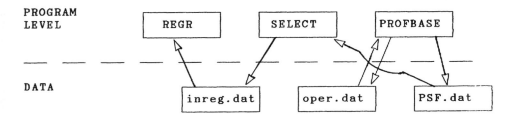

PROFBASE

This program level creates and maintains the data bases.

SELECT

This program level sorts out the operation types which shall be included in the regression analysis.

REGR

The REGR level is used for performing the regression analysis, and calculate operation error rate for user defined operations.

PSF.dat

Each record in this datafile includes data with respect to a failure rate test for a certain type operation, (eventually similar expert judgement), including PSF values, failure rate, information source, etc.

To distinguish between operations, these can be classified in two levels.

A letter identifies operation type   i.e. H
A figure identifies operation subtype   i.e. 12

The following operations are defined in the present database as shown on the following printout of the PROF user dialogue:

```
        Selected operation types:                        P R O F
                                              Copyright  A/S Quasar Consultants
   Database        Select operation types    Human reliability          Quit
   Select operations to include in regression analyses.
   ══════════════════════ SELECTION LIST OF OPERATION TYPES ═══════════════════
  ┌──────────────────────────────────────────────────────────────────────────┐
  │  Type   Subtype   Identification                                          │
  │                                                                           │
  │   A        0       Check reading and quantitative reading of displays     │
  │   B        0       Reading labels                                         │
  │   C        0       Inspection on a change of shifts or plant walk through │
  │   D        0       Selection of manual controls                           │
  │   E        0       Actuation of manual controls - commission              │
  │   F        0       Responding to annunciated legend lights                │
  │   G        0       Completing a set of multiple tasks                     │
  │   H        0       Communication tasks                                    │
  │           12       Record information in log book                         │
  │           13       Record data via telephone                              │
  │   I        0       Maintenance and callibration tasks                     │
  │           22       Calibrate dial by adjusting potentiometer              │
  └──────────────────────────────────────────────────────────────────────────┘
  -Move cursor    Return-Mark/Unmark operation types      F10-Quit
```

The sequence of records with respect to operation types in the data base is arbritrary. New data are added sequentially, by extending the data base.

Oper.dat

In this data base, a definition of operation type and subtype according to the operation classification in the PSF.dat file is included.

Inreg.dat

The file is used for intermediate data storage, as input to "REGR". The user may create his own files, if convenient.

CALCULATION OF ERROR RATE

It has to be decided which operation type data sets are to be included in the regression analysis. Taking the operation communication task as an example, the operation for which the error rate shall be calculated, could be such that all communication tasks are included. The operation also could be such that only telephone communication should be included. The more records that are included, the more precise will the data base for the statistical regression analysis be. On the other hand, there has to be reasonable correspondance with the operation in question.

When it has been decided which operations and suboperations are to be included, the program level "SELECT" is run. The user gives operation types and subtypes as input. The data base PSF.dat is read sequentially, and records with type/subtype operation classifications as defined, are picked out for the regression analysis, and the weights and the constant are calculated as shown on the following printout of the PROF user dialogue.

```
        Selected operation types:                          P R O F
  HO
                                             Copyright  A/S Quasar Consultants
  Database        Select operation types     Human reliability          Quit
  Perform the analysis.
  ════════════════════════════════ CALCULATED WEIGHTS ════════════════════════
  ┌─────────────────────────────────┐
  │ Regression analyses results :   │
  │                                 │
  │   Weight    1      0.21         │
  │   Weight    2      0.17         │
  │   Weight    3      0.07         │      ┌──────────────────────────────┐
  │   Weight    4      0.16         │      │ Number of psfs          :   9│
  │   Weight    5      0.21         │      │ Number of observations  :  24│
  │   Weight    6      0.25         │      │ Unknown psfs            :   0│
  │   Weight    7      0.31         │      │ Number of equations     :  24│
  │   Weight    8      0.33         │      │ Unknown parameters      :  10│
  │   Weight    9      0.34         │      └──────────────────────────────┘
  │                                 │
  │   Constant       -8.3625        │
  └─────────────────────────────────┘

  F1-Help                     Return-Continue                        F10-Quit
```

To calculate error rate for operations, the user has to specify PSF-values for the operation in question. In order to avoid systematic errors, the user has to interpret the PSF scale consistent with the scale applied for the data in the data base.

The sequence is as follows:

Having selected the type of operation, one can then start to feed the PSF values into the program as shown on the following printout of the PROF user dialogue.

```
       Selected operation types:                      P R O F
   HO
                                          Copyright  A/S Quasar Consultants
   Database        Select operation types  Human reliability          Quit
   Perform the analysis.
   ================== HUMAN RELIABILITY ESTIMATION ==================
   ┌──────────────────────────────────────────────────────────────┐
   │   Name of operation:  TEST                                    │
   │                                                              │
   │        ┌─────────────────────────────────────────────┐       │
   │        │ Fill in PSF values (1 -10):                 │       │
   │        │                                             │       │
   │        │ PSF-1 Complexity of the operation.........: 3.0     │
   │        │ PSF-2 Information quality.................: 3.0     │
   │        │ PSF-3 Feedback............................: 3.0     │
   │        │ PSF-4 Breaking of stereotypes.............: 3.0     │
   │        │ PSF-5 Distractions .......................: 3.0     │
   │        │ PSF-6 Time pressure.......................: 3.0     │
   │        │ PSF-7 Exposure to danger..................: 3.0     │
   │        │ PSF-8 Task relevant experience ...........: 3.0     │
   │        │ PSF-9 Personnel qualifications ...........: 3.0     │
   │        └─────────────────────────────────────────────┘       │
   └──────────────────────────────────────────────────────────────┘
   -Move cursor      Return-Select edit field    F10-Continue(Quit)
```

The program then estimates the operator failure rate, and the outcome is given like shown on the following printout of the PROF user dialogue:

```
       Selected operation types:                      P R O F
   HO
                                          Copyright  A/S Quasar Consultants
   Database        Select operation types  Human reliability          Quit
   Perform the analysis.
   ================== HUMAN RELIABILITY ESTIMATION ==================
   ┌──────────────────────────────────────────────────────────────┐
   │   Name of operation:  TEST                       Weight   Value │
   │      PSF-1 Complexity of the operation.........:  0.21    3.0   │
   │      PSF-2 Information quality.................:  0.17    3.0   │
   │      PSF-3 Feedback............................:  0.07    3.0   │
   │      PSF-4 Breaking of stereotypes.............:  0.16    3.0   │
   │      PSF-5 Distractions .......................:  0.21    3.0   │
   │      PSF-6 Time pressure.......................:  0.25    3.0   │
   │      PSF-7 Exposure to danger..................:  0.31    3.0   │
   │      PSF-8 Task relevant experience ...........:  0.33    3.0   │
   │      PSF-9 Personnel qualifications ...........:  0.34    3.0   │
   │                                                              │
   │      Constant: -8.3625      Estimated human failure rate: 0.0057│
   │      Log F   : -2.2459                                        │
   └──────────────────────────────────────────────────────────────┘
   F1-Help                                                  F10-Quit
```

FURTHER DEVELOPMENT OF PROF

The ongoing project in 1988 between STATOIL, Norsk Hydro and QUASAR is concentrating on the following development of PROF:

1. Extension of the database

FURTHER DEVELOPMENT OF PROF

The ongoing project in 1988 between STATOIL, Norsk Hydro and QUASAR is concentrating on the following development of PROF:

1. Extension of the database

   The extension will comprise gathering/registration of data from simulator tests and experience data from control rooms in operational plants, etc.

2. Systematisation of PSF's and operation types

   A more precise estimation of the PSF's and the relationship between operation types and PSF's will be performed.

3. Statistical evaluation of estimated human reliability.

4. Program user friendliness improvments.

   With this implementation, PROF should be a valuable tool regarding:

1. Estimation of human error rates when performing risk anlaysis.

2. Construction/development of information systems/control systems.

3. Control room development.

REFERENCES

1. Swain, A.D. & Guttman, H.E.
    Handbook of Human Reliability Analysis with Emphasis on Nuclear Power Plant Applications. NUREG/CR-1278 1984.

2. J.G. DeSteese, et.al.
    Human Factors Affecting the Reliability and Safety of LNG Facilities: Control Panel Design Enhancement. GRI-81/0106. 1. January 1983.

3. A/S Quasar Consultants
    User Guide to PROF. June 1987.

NEW DIRECTIONS IN QUALITATIVE MODELLING

Deborah Lucas and David Embrey,
Human Reliability Associates Ltd.,
1 School House, Higher Lane,
Dalton, Wigan,
Lancashire, WN8 7RP, UK

ABSTRACT

Qualitative modelling refers to the identification of those human interactions with a system which are likely to give rise to a human error with potentially serious consequences. The aim of qualitative modelling is to specify what actions, goals, and plans a human agent, actor or operator is likely to do or have under a given set of circumstances. Such modelling may be divided into the identification of possible error modes, together with their causes and contributing factors and the representation of such predictions for the chosen purpose of the analyst.

There are three main functions of qualitative modelling:-

a. In accident investigation, where the cause of a human error must be established so that effective error reduction strategies may be devised and implemented.

b. In the design of new products and systems so that man-machine mismatches which are likely to lead to errors are identified and eliminated before the design cycle is complete.

c. In risk assessment, where the prediction of human intentions and actions which may initiate, compound or resolve a critical system state is vital.

A number of qualitative modelling techniques have already been developed in response to one or more of these functions. Although it is not the purpose of this paper to review each of these techniques it may be said that they are, in general, limited for a number of reasons:-

- They concentrate on the representation of errors and provide virtually no assistance in identifying human failures.

- They tend to be oriented towards risk assessment and the quantitative prediction of human error rates. Factors such as the production of error reduction strategies receive very little attention.

- They are not built using a cognitive (information processing) model of the human. However, it has been shown that considerable insights into the nature of human error may be gained through such models.

New approaches to qualitative modelling are now emerging which promise to improve the capability to predict human errors together with their associated causes. The new tools are based on cognitive models of human information processing and use the new developments in artificial intelligence (AI) and expert systems modelling. This paper will describe a number of such methods and illustrate their utility and potential areas of application.

MANAGEMENT FACTORS AND SYSTEM SAFETY

DR SUSAN WHALLEY and DR DAVID LIHOU
Lihou Loss Prevention Services Ltd
Grays Court, 1 Nursery Rd, Edgbaston, Birmingham, B15 3JX

ABSTRACT

The Manager of the 1980's is now under threat of personal accountability for safety decisions; note the enquiries following the Chernobyl and Zeebrugge incidents. This means that the successful Manager (Director) now has the direct and personal incentive to consider all aspects of the system and job which may influence safety.

System goals must be achieved whilst maintaining safety, yet there are aspects of the environment which will influence the individual manager's success. These influences form two types; **firstly** the manager's influence on sub-ordinates and **secondly** the influences of the system upon the manager. Such factors as communication and motivation are recognised influences [1]; other factors are highlighted in this paper with some specific examples.

Two techniques are briefly introduced, MORT and Statement analysis, which can be used to analyse the success of management influence on safety.

The fundamental goals of this paper are to enhance awareness of managerial influence on safety and to encourage individual managers to evaluate their loss prevention effort.

INTRODUCTION

Managements understand the need to balance the *cost* of additional safety effort against the *savings* realised by loss prevention. The penalty for ineffective safety management can sometimes appear remote to individual managers who perceive a greater need to make profits for their company than to concern themselves with preventing something which has never happened and *may* never happen on their plant. The realities of *probability* and *risk* need to be experienced to be appreciated.

Public Accountability
Despite the intent of the Health and Safety at Work Act 1974, only within the last two years has there been a perceptible change in emphasis in terms of individual *accountability* for accidents.

Public concern is heightened after each disaster. Courts of enquiry are now more likely to apportion blame on directors as well as line managers and individuals as the 1987 Zeebrugge ferry disaster showed. Similarly, the USSR penalised people at all levels who were shown to be responsible for the 1986 Chernobyl disaster.

In addition to this more direct personal accountability of individual managers for safety, changes in the Consumer Protection Act mean that there is no longer a requirement of the user who suffers injury to prove that the manufacturer was negligent, providing that it is established that the given product caused the injury. Insurance companies expect to have to increase certain premiums by 50 to 100% in order to indemnify companies against these types of claims. The ability of companies to demonstrate effective safety management should attract lower premiums and enable them to limit this increase in fixed costs.

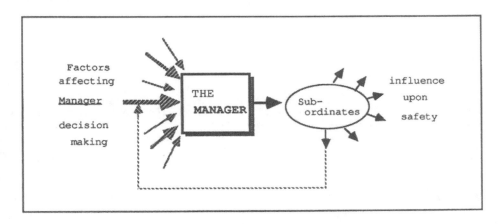

Figure 1. The Manager's position within the safety chain

174

MANAGEMENT INFLUENCE ON SAFETY

Suokas [2] identified the following eight characteristics of companies having low accident rates:

1. Management commitment to and involvement in a specific safety programme plus general plant safety matters.

2. A humanistic approach to dealing with employees with an emphasis on frequent positive contact and interaction.

3. Good employee selection procedures

4. Frequent use of supervisors to train employees

5. An emphasis on good housekeeping and general plant cleanliness

6. Attention to the qualities of the plant environment (low noise and heat, good ventilation and lighting).

7. General availability and use of personal protective equipment as and when required

8. Low staff turnover and absenteeism leading to a more stable workforce

In other words, for safe operations, management must be an integral part of the team and safety should be included as a *general operating principle*. Safety needs to be viewed as implicit within normal operating and maintenance activities, project organisation, research and design etc. In addition the concept must be supported by a trained Safety Department headed by the Safety Adviser.

Figure 2 provides a summary of managerial influences that appear to have an effect on safety. These can be divided into five key areas; general Management(company) Style, Communications, management of Personnel, management of Resources and the individual Personalities of managers. It is suggested that the direct extent of influence is roughly of the proportions shown (this suggestion is based on the authors' experience and knowledge of the literature). The two key factors appear to be *management style* (nb Chernobyl, Flixborough, Three Mile Island, Bhopal) and *communications* (nb Bhopal, Aberfann, Zeebrugge)

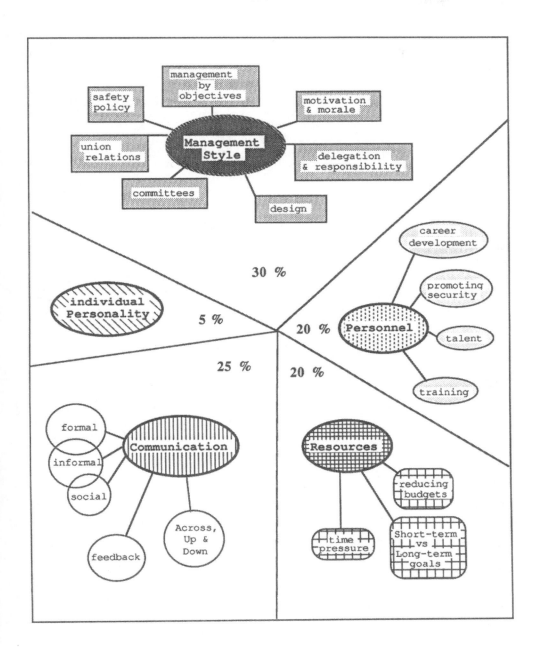

Figure 2. A summary of managerial influences on safety

Management Style

It is easier to describe the benefits of effective management of system safety than the individual roles performed by managers and even more difficult to measure the effect of management style on loss prevention.

The most obvious starting point for examining the influence of senior management on company safety is their commitment to the *Company Safety Policy* . It is important that not only line management but the General Manager regularly visits the "shop floor" to gain first hand feedback from their employees about how to improve the effectiveness of the *company's safety effort*. Safety Committee meetings do offer a forum for discussing safety matters, but these are formal contacts. Plant visits offer more scope for busy senior management to influence safety attitude in their company.

Senior managers comprising the *Operating Committee* of a company must work as a robust team to resist allowing production and financial *pressures* to undermine their loss prevention objectives. If the Production Director and the Engineering Director are continually blaming the other departments for these *pressures* and the responsibility for safety rests with the Personnel Director, one may find that plant managers and their subordinates merely pay lip service to loss prevention and that equipment is operated under crisis conditions with a policy of *breakdown maintenance*. An effective way of protecting loss prevention objectives from these *pressures* is to have a Company Safety Adviser on the *Operating Committee* who is accountable directly to the General Manager for the effective pursuit of the *Company Safety Policy* .

Senior management and the Safety Adviser must understand, measure, and if necessary, modify the factors affecting the decision making mechanism of their line managers in respect to loss prevention.

Communications

A company which relies too heavily on formal documented memoranda and proformas for recording events may find that people do not derive the maximum benefit from the information which is contained within these documents; they are by necessity reading them and therefore unlikely to discuss them with colleagues. An alternative would be to encourage regular meetings to discuss a wide range of factors relating to the safe operation of the plant. In this respect the hazard workshop type of meeting which was developed in Imperial Chemical Industries is an excellent forum for the exchange of safety related experiences. Properly structured these meetings provide invaluable ways of determining what people know, what they understand and their overall perception of the relevance of procedures. They provide an opportunity to determine whether procedures and formal communications are becoming cumbersome. Hazard workshops present the oportunity to discuss accidents and incidents which have happened elsewhere. In this environment people are more relaxed about discussing problems which they are having on the plant and drawing analogies with other peoples' problems.

Social contact outside the working environment fostered by an active social club, sporting teams, cocktail parties at which a wide range of people from the company are invited, all form a vital link in establishing an 'esprit de corps' and can be important forms of communication about what is happening on plant but more importantly this tends to create a team spirit which allows people to work together in emergencies and when there are difficult situations developing on plant. An additional benefit of social clubs is that often local residents form a major element of the club membership. This means that through the social club it is possible to communicate to the community the true nature of plant operations and probably allay any unfounded fears regarding the possible impact of the plant on the community.

METHODS FOR ASSESSING MANAGEMENT EFFECT ON SAFETY

Both a major emergency and a minor accident have two distinct phases that management can influence:

1. The cause of the incident – related to Accident Prevention

2. The effectiveness of the response – related to Loss Prevention and Emergency response

Management Oversight Risk Trees (MORT)

Accidents typically result from a chain of contributory events and influencing factors. One technique that can be used to assess the contribution of inadequate management to an accident is MORT [3].

The first part of this assessment produces an incident report in the form of an event chain. This details the direct or obvious sequence of events which is expended by adding the contributory factors, secondary events and conditions, and finally by including the systemic factors (aspects dependent upon fundamental company policy) for example; the level of supervision, procedures, training and system design.

Once the accident causes have been established, the assessor is recommended to complete the primary analysis using the MORT chart. This chart follows a tree construct similar to fault tree analysis and adopting the same logic, ie AND/OR gates to link the branches together. Each branch represents a particular responsibility of management, with associated questions to establish the current management adequacy – each aspect must be assessed as Adequate or Less Than Adequate with the evidence recorded as to why. The assessor is forced into making a decision.

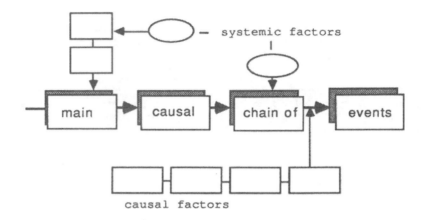

Figure 3. A typical MORT accident event chain

Note that it is not necessary to have experienced an accident within the company before using MORT, since the charts can be used from the bottom up to establish areas of management weakness which could lead to an accident. In other words it can be used to highlight where improvements are required before any problems occur.

The general structure of the main MORT charts commences with a split between oversights and omissions versus assumed risks. Only those risks which have been identified, analysed and accepted at the appropriate management level can be deemed to be assumed risks. It is important to remember that mistakes could have been made when initially accepting a risk, therefore the assessment should still be applied.

The next major sub-division separates 'what happened' from 'why'. The 'what happened' considers the specific control factors that should have been in operation, whilst the 'why' considers general management system factors. It is the 'what happened' branch of the tree which forms the major assessment route during an accident analysis, whilst management system factors are of prime importance if assessing the company's safety program.

Statement Analysis
A second technique that can be used to review an incident is that of statement analysis. Statements, or more generally information, can be *grouped into sets* relating to the cause of the incident (or break-down in accident prevention), the development of the incident and the remedial actions. These different groups of 'statements' can then be analysed for *influencing factors*.

This method of assessment was carried out for the Bhopal incident [4] with statements grouped as follows:-

1. **Company Profile** ie. the corporate influence on the Bhopal plant
2. **Procedures** ie. current operational practice just prior to the incident
3. **Incident Development**
4. **Emergency Response** by both the company and the authorities
5. **Regulatory Controls** both before and after the disaster

The associated influencing factors (Performance Shaping Factors) check list is also provided in Table 1. In terms of possible management influence on the Bhopal disaster the following PSFs must be considered:

Corporate Factors – CF1, CF2, CF3, CF4, CF5, CF6

Personal Interface – PI1, PI2, PI3

Environmental Factor EF1

Equipment Factors – EQ1, EQ3

Personnel Factors – ID1 (from a total = 31)

However it would be an equally logical stance to suggest that *all* factors could have been influenced by general management decisions.

From the information available relating to Bhopal the three most frequently implicated PSF groups were as listed below with their frequency shown as a percentage of the total number of PSFs identified.

CF: Corporate Factors 33.1%

PF: Personal Factors 19.2%

PI: Personal Interface 14.6%

Note that **CF** and **PI** groups are extensively influenced by management

It was also important to note that the two dominant *statement groups* in terms of the proportion of associated PSFs were Company Profile (29.8%) and Procedures (29.1%).

Clearly Statement Analysis is only as precise as the information that it is based upon; if certain relevant aspects are not contained within the available information they will not be assessed.

Table 1
Performance Shaping Factors

Corporate Factors

CF1.   Technology: - process/equipment/computers

CF2.   Personnel: - management/supervision/workers/
engineers/scientists/experience/training

CF3.   Headquarters: - on site/elsewhere/remote/on call/
financial influence/technical influence

CF4.   Technical Support: - R&D/quality control/pilot
plants/specialist equipment/technical experts

CF5.   Communications: - formal/personalised/specific/
general/technical/oral/written/periodic/updating

CF6.   Procedures: - audits/surveys/systems of work/
effective/straightforward/technical reviews/
safety reviews

Process Factors

PF1.   Personnel Involvement:- continuous/continual/
occasional

PF2.   Technology:- manual/semi-automatic/microprocessor

PF3.   Chemistry: - (batch/continuous)/sensitivity/
precision required/duration novelty/predicability

PF4.   Materials: - hazards/personal contact/proximity/
variety

Machine Interface Factors

MI1.   Controls: - (direct/indirect)/location/
identification/feedback/prompts/response time/
access/visibility

MI2.   Displays:- type/location/identification/continuity
continuity/legibility/clarity/response time/
access/visibility

MI3.   Indicators:- warning/status/display & control
interactions

Personal Interface

PI1.   Communications:-(formal/informal)/direction
(sender/receiver)/method (direct/indirect)/
feedback

PI2. Information: - access/type/format/clarity/legibility/
      recording mechanism

PI3. Management:- supervision/strateg/selection/incentives/
      instructions/team work/attitude

Environmental Factors

EF1. Work Pattern: - shift work/hrs per week/overtime/
      rest periods/self regulation

EF2. Physical: - lighting/glare/atmosphere(fumes/
      gases/dust/noise/temperature(heat/cold)/vibration/
      skin irritants

EF3. Access: - space/height/stairs/ladder/terrain

EF4. Location: - control room/on plant/ plant & control
      room/various/climatic exposure

Equipment Factors

EQ1. Clothing: - standard/special/safety)/availability

EQ2. Tools:- availability/servicability/requirement/type

EQ3. System Equipment: - safety/standard/fail safe/
      backup/operator interaction/size/identification

EQ4. Operator Equipment: - furniture/domestic/utility

Demand Factors

DF1. Physical: - posture/movement/control/power/static/
      co-ordination/dexterity/vision/touch/hearing/smell

DF2. Psychological:-responsibility/concentration/memory/
      decisions/urgency/accuracy/alertness/flexibility/
      risk perception/orgainistion/monitorg

Idiosyncratic Factors

ID1. Training: - general/safety/specific/retraining/
      simulations/structured/apprenticeship/
      qualifications/amount/on the job

ID2. Experience: - amount/relevance/interrupted/
      specific

ID3. Knowledge: - completeness/accuracy/relevance

ID4. Personality: - temperament/motivation/risk taking/
      capability/sociability/mood/confidence

ID5   Health: - recent illness/fitness

Two aspects of Bhopal that can be traced directly to management decision is the change in staffing that had occurred between the original compliment allocated to the plant and the time of the disaster and the problems associated with communications during the emergency response.

Staffing Levels
Originally   2 maintenance supervisors
Time of accident: 1 maintenance supervisor

Originally   10 maintenance personnel
Time of accident: 5 maintenance personnel

Originally   2 operations superintendents
Time of accident: 1 superintendent

Originally   3 plant supervisors
Time of accident: 1 plant supervisor

Originally   12 plant operators
Time of accident   6 plant operators

Replacements
External *audits* by Union Carbide men from the USA
replaced by ----->   internal *reports* by local people

American trained safety staff
replaced by ----->   inexperienced technicians

Chemical Engineer plant manager
replaced by ----->   electrical Engineer

In terms of communications, many problems arose during the emergency response. Note that it is still unclear why the plant did not notify the factory manager directly that a release had taken place, it was left to others to make the contact. It is also important to note that there was no pre-arranged plan for notifying neighbouring factories nor for rapidly starting co-ordinated evacuations.

Bhopal like many other major disasters demonstrated the influence of inadequate management *decision making*, procedures which produced latent accident prone conditions and inadequate emergency response, allowing an accident to escalate into a disaster. Procrastination throughout the management tree is one of the most common causes of these failures.

FACTORS INFLUENCING MANAGERS

In addition to their influence on others, it is important for individual managers to be aware of what aspects can be influencing themselves and hence the success of their decision making. As people, managers are as susceptible to the negative influences of their general 'environment' as shop-floor workers.

By providing an awareness of these negative aspects, managers can be helped to recognise the conditions which degrade their performance.

The four main groups of influences upon management are briefly defined in Table 2 and a sample of these are explored in the following sections.

External Pressures Upon the Company

Regulations. In England, companies are overseen by Government Safety Inspectorates and have to conform to certain legislation; eg the health and safety at work act, and the factories act. In addition there are published guidelines linked to safety aspects; eg the British Standards, Health and Safety Executive Guidelines. Obviously if such regulations and advisory documents exist it is possible for a company to keep copies for individual managers to refer to. If such documents are not available the individual manager must realise that any decision making, in particular in terms of the implications for safety aspects, may be lowered in quality of judgement

Table 2

Factors influencing management

A. *External pressures upon the Company*

1.	Regulations –	the existance of inspectorate, law and safety guidelines
2.	Market economy –	state of the country and performance of the specific industry
3.	Cultural norms –	population expectations and attitudes

B. *The Company*

1. Organisation character –
    structure, size and formality of control
2. Management level –
    management structure and the chain of command
3. Individual's role –
    inter-departmental or specific and isolated
4. Environment, Rewards and Motivation –
    working conditions and incentives, their effect on creating loyalty to the company
5. Interactions and communications –
    the existence of communication channels and the promotion of communication
6. Informal influences –
    the organisations sub-culture, internal politics, power struggle and prestige, allocation of resources

C. *The individual Manager's Job*

1. Responsibilities - direct or indirect inclusion of safety, knowledge of boundaries of responsibility and those responsible for other aspects
2. Functions - knowledge of requirements, their differentiation and planning
3. Tasks - variety and cohesion

D. *The individual Manager's Personality*

1. Management Style - the ability to deal with stress, delegation versus autonomy
2. Time allocation - time management, identifying needs in advance of crisis, balancing different demands
3. Judgements - recognition of important issues and appropriate response

Market Economy. This broad title covers a multitude of factors that may tempt the manager into poor safety assessments or ignoring safety implications. Such aspects as diverse as a generally depressed economy, reduction in percent of the market, new alternative products, increases in raw material costs, falling profits, tightening of department budget, public attitude, take-overs and mergers all affect the individual managers budget and hence any allocation of resources. If the company is doing well then individual managers can look further than short-term goals and performance and consider long term implications and possible consequences of decision making. Conversely, if the company is having financial problems cut-backs are inevitable. In this situation, it is easy for the individual to place safety aspects as a very low priority. However, if the particular process is high risk, specifically in terms of potential consequences of accidents, then the urge to take the risk must be suppressed. The safety implications of decision making cannot afford to be ignored (as much for the sake of the company as well as the community - litigation may bankrupt!)

Cultural Norms. Expectations and generally accepted attitudes towards operations and safety are obvious influences on an individual manager. If government and the population demand a high standard of safety those within the company will adopt these standards (eg. Sweden). If the overriding concern is production, minimising expenditure and attracting industry to the area then safety will be given a lower priority (eg. Third World countries). Managers in Third World countries must make a more conscious effort to maintain safety standards to compensate for less stringent and fewer Regulations.

If the national characteristic is to be methodical and thorough safety aspects are unlikely to be overlooked by the management. Conversely if there is a national tendency towards risk taking the maintenance of safety standards will depend upon the professional values of individual managers.

Company Pressures upon its Managers

Organisation Character (size, structure,type). The larger the company, the less responsible individual managers may perceive themselves to be. There will be links with role allocation such that if roles are poorly defined then each manager may consider that ensuring safety is someone elses job. Therefore, in particularly within large companies, it is important that all managers consider safety to be their own responsibility. That said, the company must ensure that those individuals who manage with safety as a primary concern are not penalised, it is suggested that this problem of negative rewards could be more prevalent within medium sized organisations.

The structure of the company may affect safety, eg the provision of a particular unit responsible for safety does not automatically create a safer management climate. The department may be given responsibility for safety but more importantly has the group been given authority? Responsibility without authority leads to lowering of respect and hence a lowering of morale for these managers, which may in turn lead to a lowering of standards. As a manager caught up in such a system it is important to document where authority begins and ends so that it is possible to demonstrate what the limitations are and what effect this will have. If the safety group which is responsible for safety audits demonstrates a problem area but has not the authority to ensure that it is rectified (preferably indirectly by support of production managers), an accident-prone environment will start to grow.

Companies that are structured upon a strict hierarchy of decision making and control may reduce communication channels and feedback, creating the situation where shop floor safety problems cannot reach those managers in the position to rectify them. In particular, where departmental segregation emanates from board level interdepartmental exchanges will be formal and rivalry may occur with success often linked to profits and budgets. Management isolation can also lead to a lack of appreciation and understanding of problems that cause indirect or non-immediate effects on the company (eg. aspects related to safety), these can be viewed as over-emphasised or non-urgent and hence forgotten or ignored (NB the request by Ships Masters for bow door indication - Herald of Free Enterprise disaster).

Managers who work in isolated positions, either in terms of other departments or levels in the hierarchy, need to recognise that they may be out of touch with the people affected by their decisions. They must endeavour to develop communication channels with these people in order to obtain the information needed to make correct decisions. Managers in this situation need to consider carefully shop-floor complaints related to safety in order to correctly estimate the urgency and potential severity of the situation. If they are in any doubt they must make a personal visit to obtain the facts before arriving at a decision.

In terms of overall responsibility for a company there are potential areas of weakness no matter which control strategy operates. Board Control with no devolution of responsibility can lead to problems of 'group think' : It has been shown that groups particularly those serving well-established conservative organisations take on a character beyond that of the individuals, suppressing individual reservations or creativity, lateral thought rarely occurs and solutions tend to repeat those that have been made before. If the group is made aware of these potential pitfalls they may be overcome. Hence improvements are difficult to make and responses to new problems may not be apt.

Private owners with no devolution of control may not be personally qualified to respond to all the difficulties posed, hence decisions may be lowered in integrity due to lack of knowledge, experience, time, technical support etc. Obviously the smaller the company the more vulnerable it is to these problems due to a lack of potential advisors. Board control with devolution is more vulnerable to "passing-the-buck" than single owner with devolution but both must ensure individual and collective responsibilities are allocated, recorded and monitored. Individual managers within these systems may operate more reliably if they request details of their functional responsibilities, where these overlap with others and who is responsible for aspects outside their direct control.

Informal Influences (internal politics, power struggles, prestige and budgets). Pressures upon managers often come from an unexpected quarter, they are not associated with the work itself but from informal influences through colleagues and peer groups.

The larger the organisation the more breeding ground for the effects of politics, power and prestige. These all link, however loosely, with individual motivation - obviously some people are more prone to these effects than others. It is important to maintain a balance within the individual between personal needs and company goals. If the individual is not receiving formal recognition and rewards, the tendency is to look elsewhere for motivation and hence the manager is vulnerable to informal pressures. Strength of personality is often an additional factor. Those who are unsure of themselves, needing external praise and recognition plus those who are low achievers but require power to feel fulfilled, are likely candidates for entering into these office games. Within organisations where this sub-culture has obtained a stronghold, there is likely to be insecurity influencing decision making, with some individual managers falling prey to back-stabbing and personality clash. The only chance such an organisation has to regain control is to recognise the situation then counter it through formal rewards (promotional, financial) restructuring and recognition through informed formal bestowal of power. Within the informal power hierarchy an empathy towards safety is rarely seen as a positive status. If the company believes in a safe system of work it must counteract this force.

The pressure of budgetting is due to demonstrating that money is well spent and success is often related to profits. As an individual manager concerned with safety, it should be possible to demonstrate the positive value of budget spent on safety. This can be in terms of rectifying problems that have in the past had safety implications or lost production, demonstrating the cost of these incidents in relation to the cost of change. Alternatively, outline possible incident scenarios if the changes are not made and the likely cost of these to the company, compared with the cost of prevention.

TO CONCLUDE

This paper has provided an overview of the need for individual managers to face their safety responsibility. It has also drawn out the differences between the influence management has upon safety and those factors that can impair management decision making with respect to safety. In terms of management effect upon safety, two techniques have been demonstrated that can be used to identify either the contribution management has made to an accident or the current standard of management structures within the organisation.

Many aspects of the management *of* safety and management contribution *to* safety have been omitted for brevity. What has been included is sufficient to demonstrate that 'blame' or more positively, the causal analysis of accidents is no longer confined to the 'point of no return' ie the final action that can be directly attributed to the start of the incident or disaster. Causal chains can be traced all the way up to top management. Their decisions influence the initial design of a process or system, the maintenance expenditure and programme, communication structures and response to interim safety problems, accident mitigation, containment or emergency response. All these aspects and many more contribute to the existance of an 'accident-likely' environment. Such early errors of judgement have been termed 'latent errors' [7] to distinguish them from the 'active errors' that immediately precipitate an incident.

Individual managers, as well as a company's management as a whole, will benefit from an appreciation of their direct and personal influence upon their company's safety and existence. Acknowledgement of a problem or a potential problem is the first step towards solving it.

REFERENCES

1. Bellamy, L., Neglected individual, social and organisational factors in human reliability assessment. _4th National Reliability Conference - Reliability '83_, 1983, pp. 2B/5/1 - 2B/5/4

2. Suokas, J., The role of management in accident prevention. _1st International Congress on Industrial Engineering and Management_, Paris, June 1986.

3. Johnson, W.G., _MORT Safety Assurance Systems_, Marcel Dekker New York, 1980.

4. Lihou, D.A. and Whalley, S.P., Bhopal- some human factor considerations. _World Conference on Chemical Accidents_, Rome, CEP Consultants Ltd, 1987, pp 88-92.

5. Reason, J., Summaries of 5 case studies illustrating the part played by latent failures in accident aetiology. Department of Psychology, University of Manchester, 1988.

189

ADDRESSING HUMAN FACTORS ISSUES IN THE SAFE DESIGN AND OPERATION OF COMPUTER CONTROLLED PROCESS SYSTEMS

LINDA J. BELLAMY and TIM A. W. GEYER
Technica Ltd, Lynton House, 7/12 Tavistock Square, London WC1H 9LT

ABSTRACT

Incidents that occur in computer controlled process systems would appear to involve human error at all stages from design through to operation. Some examples are given. To overcome the sources of error is problematic because guidelines and analytic methods specifically relating to human factors in computerised process control do not exist.

We have attempted to address some of these problems by highlighting areas that could be considered in providing design guidance and by emphasising the need for a design review methodology.

INTRODUCTION

Work has been carried out by Technica for the Dutch authorities to develop a methodology to review the design, operation and modification of computer controlled process systems. As part of this work, human factors issues were considered, particularly those aspects relating to safety.

As far as we know, there are no accepted well defined design standards or methodologies for dealing specifically with the human component in computerised process control systems. The recent PES (Programmable Electronic Systems) Guidelines (HSE, 1987) identifies the importance of the operator's role, the man-machine interface, supervision, training and procedures but hardly goes into detail on these issues, and neither does it devote a special section to them. The Guide to Reducing Human Error in Process Operation (SRD, 1985), while addressing human factors issues explicitly, provides a list of guidance principles that are short, simple and concise.

This is not meant to be a criticism. Rather, it emphasises the fact that, for complex control systems, detailed point by point instruction on all aspects of design and operation would be difficult to achieve. Marshall et al (1987), in writing about guidelines for the design of the user interface for complex computing systems, say that:

"... user-interface design guidelines cannot provide an automatic solution to the design problem. They do not tell the designer how to do exactly the right thing and they do not tell him or her exactly when to do it."

They go on to suggest that, in order to make genuinely useful statements guidelines must be context free. Guidelines are often based on informed opinion rather than on hard data. They should therefore be viewed as an informal collection of suggestions, rather than as a distilled science. Designers are likely to have to make some choices of their own and be prepared to test their work empirically.

In addressing the human factors issues involved in the safe design of computer controlled process systems, it is in this spirit that this paper is written.

To put the human factor into context, a summary of an analysis of 17 accidents is first given. A simple information processing model is used to illustrate human factors thinking in carrying out such an analysis. This model then provides a basis for consideration of the kinds of human deviations that could be included in a HAZOP. Before this, a simple set of design principles for computer controlled process systems are presented.

INCIDENTS IN COMPUTER CONTROLLED SYSTEMS

Descriptions of 17 incidents which had occurred in computer controlled systems were reviewed to identify broad classes of failures. These failures led mainly to small and medium sized releases, in one case plant damage, and in another a fireball. Table 1 shows a summary checklist indicating the number of failures in each category (hardware, software, human etc.). For any particular incident it should be noted that the failures are not independent eg. some failures led to others.

From this summary it can be seen that human errors during operations were associated with 59% of the incidents. Errors were mainly due to inadequate, insufficient or incorrect information supplied to the operators (59% of incidents) and a failure to correctly follow procedures (47%). Human errors in design were involved in 29% of incidents. Hardware and software failures were less prominent. Interestingly enough, most of the causes of failure in these computer controlled plants could equally well have occurred in conventionally controlled systems.

Figures 1 & 2 illustrate how the failures can arise in the man-machine system. The model shows the basic information processing operations of the human operator. Superimposed on this are the stages leading up to an accident. These are described after introducing the model components.

The model shows that the operator **perceives** a situation based on the information displays available. It should be stressed that the operator will act on his perception of a situation which may not always reflect the real situation. **Decision and response selection** is based on information perceived and information stored in memory. **Long term memory** refers to stored knowledge gained from experience, training etc. which influences the way we perceive new experiences. **Working memory** is a short term store for data momentarily required for a particular task (eg. remembering a telephone number long enough to dial it). This short term storage mechanism has a limited capacity. Following decision and response selection, **execution of a response** may occur.

Perception, decision making, response selection and response execution all require **attention resources.** These resources are limited and can be exceeded under adverse conditions (eg.high stress). When responses are well learned there is less need for such resources (eg. consider the development of car driving skills).

Response execution usually involves acting on some controls to affect the process, which may cause a change in displays. This forms a feedback loop whereby an operator gains information about the effects of his control actions. In a highly automated system, where the operator acts principally as a monitor and only steps in when the automatics fail, the majority of actions may involve searching displays, paging through the data base, logging values etc.

Failures can occur at any point in the model. In the examples shown on the model, the course of events can be seen by working through the numbered comments in order.

In the incident shown in Figure 1 the bottom discharge valve of a reactor was open when a batch job was started. The operator thought the valve was closed because this was the status displayed. The result was a release of more than 15 tons of vinyl chloride gas.

Figure 2 shows how problems can arise when the operator does not have all the required information available in parallel. In this incident the operator focussed all his attention on the furnace such that he missed what was happening near the scrubber. The fact that the alarm display was a scrolling screen showing only the last 12 alarms resulted in the low level alarm for the cooling system being missed. For this event where lots of alarms were being triggered, the operator lost control of exactly what went wrong and where. As a result, serious damage to part of the plant occurred due to exposure to extremely high temperatures.

Table 2 provides a more detailed breakdown of the causes of the 17 accidents. As can be seen, poor information provision, whether incorrect, hidden, or not available derives from a number of sources. It is likely in some cases that the quality of procedures, supervision and checking were insufficient to enable errors to be identified and recovered (eg. in installation and maintenance). On the other hand, over reliance on the computer when carrying out procedures could reflect inadequate understanding by operators of the functions performed by the computer and how these are carried out.

TABLE 1
Checklist used to identify broad classes of failures for 17
computerised process control system incidents (individual incidents
may be associated with more than one failure category)

FAILURE CATEGORY		NUMBER OF INCIDENTS	% OF INCIDENTS
HARDWARE	Computer Hardware	3	18
	Connection Hardware		
	- Electronic	0	0
	- Pneumatic	0	0
	- Electrical	1	6
	Protective System Hardware	0	0
	Equipment Hardware	5	29
SOFTWARE	System Software (Manufacturer's Shell)	1	6
	Site software implementation (i.e. Software written for the process plant and installed during and after implementation)	2	12
HUMAN	Error Context		
	- Design	5	29
	- Installation	2	12
	- Commissioning/Testing	1	6
	- Operating	10	59
	- Maintenance	2	12
	Error Type		
	- Failure to follow procedure (correctly)	8	47
	- Recognition failure, given adequate supply of information	2	12
	- Error due to inadequate/insufficient/incorrect information supplied to person(s) involved	10	59

TABLE 2
Breakdown of causes of the 17 incidents

HUMAN AND SOFTWARE ERRORS	INCIDENT NUMBER CODE
Interface does not display actual plant status	1, 6, 8, 16
Installation error leads to incorrect information	3, 8
Alarm set incorrectly	4
No alarm (maintenance error)	4, 5
No alarm (design)	4, 5
Operator misses information due to overload	2, 13, 15, 17
No independent means of cross checking provided	1, 3, 6, 16
Operator fails to cross check	8
Trip disabled/manual override	1, 8, 11
Over-reliance on computer	9, 11, 14, (15)
Inadequate knowledge	(3), 11
Failure to update operators' information	12, 17
Incorrect control signal (maintenance) error	10
Design error: Plant	4, 5, 8, 17
Design error: Computer control system	7
Software error	7, 9
HARDWARE FAILURES	
Equipment hardware	2, 5, 13, 14, 15
Computer hardware	7, (11), 16
Connection hardware (electrical)	6

(Note: Incident numbers in parenthesis indicates that there was not enough information to allocate to the failure category with certainty).

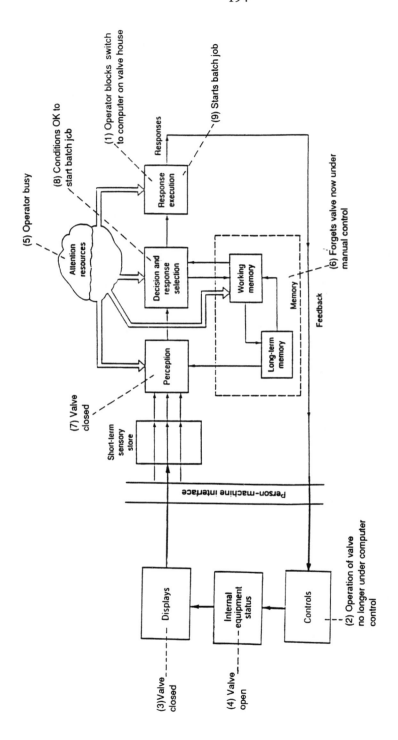

Figure 1. A model of man as an information processor in a man - machine system (adapted from Meister, 1971 and Wickens, 1984) showing analysis of incident 1

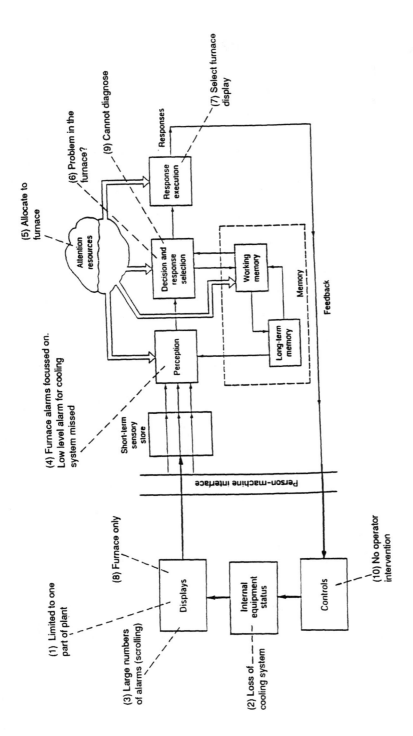

Figure 2. A model of man as an information processor in a man - machine system (adapted from Meister, 1971 and Wickens, 1984) showing analysis of incident 2

DESIGN GUIDANCE FOR THE MAN-MACHINE INTERFACE

Human-Computer Collaboration

Bainbridge (1983) has summarised some of the major problems that automation brings to the design of the operator's task, interface with the system, training and procedures. Bailey (1982) discusses some of the problems of allocation of function between the human and the machine. If one extracts the major issues applicable to computer systems from these two sources, a set of general recommendations can be made:

1. Operators should not be left with an incoherent set of functions that the designer cannot think how to automate. Operators need proper support for carrying out tasks after automation and this means thinking about how the operator and computer collaborate in carrying out the various control functions.

2. When the role of operators is mainly one of monitoring, it is essential to maintain the necessary operator skills, knowledge and mental model of the system. This can be achieved in two ways:

> (i) Allowing operators to take over from automatic operations to get "hands on" experience.
> (ii) Use of high fidelity simulators where realistic failure scenarios can be used to train operators to adopt good general problem solving strategies (eg. for low probability events) rather than specific responses such that these strategies can be used in cases of unanticipated failures.

3. It is essential that operators are aware of exactly which parts of the system are under computer control and which are in manual mode, especially in high periods of activity such as an emergency. If the operator needs to follow what the computer is doing (eg. in an automatic shutdown) it may be necessary to think about presenting this information in new ways that are compatible with his skills eg. slowing down the display of automatic events that are too fast for him to follow. If this is not possible, then one cannot allocate this role to him.

4. It is important that failures are made clear to the operators in time for them to both think out what to do as well as take corrective action. The control system should not disguise the failures to limit this thinking time. The need for operators to think out the effects of possible actions must be considered in design when selecting amongst alternative solutions.

5. Clear criteria need to be provided for the operators to indicate when there is a necessity to take over from the automatic operations as operators may not be able to work this out for themselves.

6. The relative merits of human and computer control should be taken into account. Human operators have distinct strengths above that of machines for certain tasks, e.g. adapting to a novel set of conditions, pattern recognition, etc., whereas machines are better than humans at others e.g. fast responses.

Principles of Interface Design

Interface design principles were developed for the specific problems involved in computer controlled process systems. The principles refer essentially to operator monitoring and control tasks.

There were 5 main principles used:

[A] Provide the operator only with information that he needs <u>and none he does not need</u>.

[B] All the information relating to a particular task should, as far as possible, be grouped together in one place.

[C] Operator's experience affects the way they read a display or operate a control, so their expectations should not be violated as they move from one physical location (or VDU page) to another.

[D] The design of the interface should be compatible with the operator's limitations and capacities as an information processor.

[E] Manning should meet resource requirements. Personnel should not be predominantly either overstressed or bored.

The principles were broken down into a number of specific recommendations. An example is shown below for Principle B.

Principle [B] All the information relating to a particular task should, as far as possible, be grouped together in one place.

[B1] Determine information requirements for tasks by carrying out task analyses.

[B2] Controls and displays related by action and effect (feedback) should be located together as far as possible.

    [B2.1] All the effects of a keystroke command on the process should be simultaneously observable on the operator's displays. If the process response time is slow some feedback must still be given that the action has been initiated.

    [B2.2] If more than one person must work on the same part of the system, all the relevant information should be simultaneously available to a person coordinating the task. (This includes the coordination of control room and maintenance tasks etc.)

[B3] As far as possible, supply <u>all</u> the necessary information <u>simultaneously</u> (i.e. in parallel rather than sequentially) that is needed for a diagnosis or a control decision.

    [B3.1] The operator should not have to page through the displays to collect together all the information relating to a particular failure

[B3.1.1] Sufficient VDUs should be available for simultaneous display of the required information if it is likely to appear on different display pages.

[B3.1.2] As far as possible within physical and ergonomic constraints, all the information needed for diagnosis of one failure should appear together on one display page. Therefore all the variables affecting a controlled state should be, as far as possible, displayed together.

[B3.2] The minimum number of VDUs will partly be determined by the number of unrelated failures that could occur simultaneously:

[B3.2.1] Never use only one VDU per workstation for monitoring and control tasks.

[B3.2.2] Additional VDUs may be needed for dedicated displays (e.g. alarms).

[B3.3] Certain display divisions are acceptable. These are cases where the cause and effect relationship between plant/ process variables is simple. Different display pages should not cut across interacting variables. This point relates not only to the division of displays at one operator station, but also division of displays between operator stations.

[B3.4] The operator will need to be able to see cause and effect relationships, time lags and rates of change in the process.

[B4] Minimise uncertainty.

[B4.1] Provide an overview display that will satisfy the operator's need to keep a summary check on the whole of the system for which he is responsible. (This could be a wall mounted display.)

[B4.1.1] Provide alarm overviews that are permanently on display.

[B5] Avoid operators having to move around too much to different locations to collect or transmit information.

[B5.1] Consider using flexible as well as fixed communication equipment.

[B5.1.1] Communication systems for transfer of current information should not require operators to leave their consoles.

[B5.2] Consider conference facilities if communication needs exceed one-to-one for coordinated tasks.

[B5.3] It should be possible to display any information from the plant data base on any VDU.

[B6] Centralise important information.

[B6.1] Consider using a dedicated alarm VDU at operator workstations.

[B6.2] Consider providing a summary of important information for supervisors to allow prioritising of actions.

[B7] Locate related items such that they are easy to associate.

[B7.1] Locate alarm displays close to (or on) other displays with which they are associated.

[B7.2] Group alarm summaries in a meaningful way (i.e. according to sequence, priority, function etc.)

[B7.3] Locate acknowledgement devices such that alarms cannot be acknowledged without being identified first.

[B8] Avoid two operators (or an operator and supervisor) being able to simultaneously affect the same part of a process from different VDU/keyboard locations

[B8.1] If [B8] is unavoidable, information on each operator's actions will have to be provided and supervised in these situations, imposing an additional monitoring load. (If there are two unrelated failures this may not be a problem).

We do not consider that this is necessarily a comprehensive list. However, we have endeavoured to cover all the major areas which are highlighted by previous accidents, by ergonomics analysis of the problem areas (e.g. Bainbridge 1983) and current ergonomics practice.

HUMAN FACTORS HAZOP REVIEW OF COMPUTER CONTROLLED PROCESS SYSTEMS

Many incidents arise in non-safety critical areas of the plant, as was found in the 17 that were examined here. For this reason guidelines which only address critical safety systems are insufficient where the design review needs also to cover incidents with the potential to cause damage or serious environmental effects. We have already shown that human error plays a large part in such cases.

We consider that extending hazard and operability studies (HAZOP) to include human factors could go some way towards dealing with problems in design which could lead to human errors with consequences relevant to plant safety. HAZOP is a method for checking a design by applying a limited set of guidewords and variables to examine the suitability of the design to respond to a whole range of deviations.

Deviations are derived by combining a set of guidewords (eg. NO, WRONG etc.) with a set of variables (eg. SIGNAL, ACTION etc.) and these deviations are then applied to some element of the design in the form of questions (eg. "What happens if there is no signal when there should be").

By adapting the information processing model (Figure 1 and 2) to this format we have derived the following:

	GUIDEWORDS	VARIABLES
	MORE	INFORMATION
	LESS	ACTION
	NO	
	WRONG	

The variable INFORMATION applies to information available from displays, procedures, previous training, experience, communications and any other source which an operator may use. The variable ACTION refers to the operator response. Errors in ACTION may be in terms of incorrect selection or incorrect execution of a response.

A set of specific deviations can then be provided for each of the 8 deviation categories. An example for NO ACTION is given below:

NO ACTION

This deviation occurs when the operator fails to act when there is a demand to do so.

Example Causes

Control cannot be accessed
Error recovery not possible
Necessity for action not perceived
No information to act upon
Action not possible
Assume computer control of operator function
No operator present
Operator distracted
Omit procedural step(s)
Communication failure
Action too late
Assume other person has acted
Insufficient time to complete
Fail to restore to automatic control
No supervision/checking/testing

We propose to test out this method in the future using details of a site specific interface design, procedures, and control philosophy documentation.

CONCLUSIONS

There is a wealth of human factors knowledge that could be put together in a simplified form to enable design engineers to incorporate human factors early in the design process when changes can be made at relatively little cost.

It would be useful if guidance principles and review methods could be standardised to enable them to be applied with confidence. To do this would require collaboration between human factors specialists, regulating authorities and industry.

REFERENCES

Bailey, R.W., (1982) Human Performance Engineering: A Guide for System Designers, Prentice-Hall Inc., New Jersey.

Bainbridge, L., (1983) Ironies of Automation. Automatica, Vol. 19, No. 6, pp 775-779.

Health and Safety Executive (1987) Programmable Electronic Systems in Safety Related Applications. Vol. 1: An Introductory Guide, Vol. 2: General Technical Guidelines, HMSO, London.

Marshall, C., Nelson, C., and Gardiner, M.M., (1987) Design Guidelines, pp 221-276 in Applying Cognitive Psychology to User-Interface Design. M.M. Gardiner and B. Christie (Eds) J. Wiley and Sons.

Meister, D., (1971) Human Factors: Theory and Practice, New York., Wiley.

Safety and Reliability Directorate (1985) Guide to Reducing Human Error in Process Operation. Short Version. U.K. Atomic Energy Authority, SRD R347, February 1985.

Wickens, C.D., (1984) Engineering Psychology and Human Performance. Columbus, Ohio, Charles Merrill.

MANAGEMENT IN HIGH RISK INDUSTRIES

by   Ian Alexander Watson, BSc, CEng, MIEE, FSRS

and Francis Oakes, CEng, FIEE, FBIM, FINSTD
Engineering and Management Consultant

1   INTRODUCTION

Important insights into the effect of operator performance on the reliability and safety of industrial plant have been gained over the last 20 years, but Bhopal, Challenger, Zeebrugge and Chernobyl remind us forcibly that greater awareness of the vital importance of human reliability factors is needed at senior management level. The impact of these events on industrial managers generally seems to have little effect on design and manufacture. Yet commercial viability depends as much on human reliability factors as of course do health and safety.

In recent years, intensive research has been under way in several centres of excellence in the UK and abroad. This had led to a better understanding of the causes of human error, of the decision making process, and of the many complex factors which shape human performance. Greater concern for more effective management of physical and commercial risk has stimulated the development of better techniques for operational audit and for the assessment of management quality.

It is commonly accepted by concerned professionals that human factors (HF) can have a significant impact on the safe and reliable operation of technological plant. This understanding is manifest across a variety of industries and technologies eg, chemicals, processing, nuclear power, aviation, mining, computers and so on.

Reliability considerations usually start with the system or plant description, specification and performance. Design, degradation mechanisms, operations and maintenance are taken into account together with data on random failures and more systematic types of failures from data banks eg the NCSR Data Bank (1) and specific data collection and analysis campaigns. Methods of dealing systematically with human factors/ reliability are slowly emerging and work is in process (2). The management factor has become apparent from many accident reports (3)(4) and also from work on the analysis of common mode failures (5). However there are at present no formal methods of dealing with Human Reliability Factor in Technology Management and their consideration is still entirely subjective and a matter of personal judgement. The environment of the plant affects its operation and some of the significant factors are shown in Figure 1. finally there are bound to be degrees of uncertainty associated with all the considerations which will produce an overall uncertainty, these can to a limited extent be expressed in statistical and mathematical terms.

A model which shows the interconnectedness of management, operators, plant and the tasks involved is shown in Figure 2. This was produced as a result of an analysis of industrial fatal accidents performed by SRD (6). Accident causes arising from management errors were found to be significant by comparison with the other factors shown. There is a view held by members of regulating agencies, which is supported by some data, eg aviation accident rate spread between airlines, that the variation in accident risk between good and bad management can be at least an order of magnitude.

The relationship between plant reliability, human reliability and management needs to be understood more explicitly than is now apparent. This can be demonstrated by showing specifically and analytically links between reliability analysis methods, operator tasks, human action theory and management structural analysis. Reviews of the Challenger accident report (3) a study of the Bhopal disaster (4) and Chernobyl (7) illustrate how organisation and management influence safety and reliability.

2 MANAGEMENT AWARENESS

Techniques to examine and improve human performance in hazardous operations exist. They have been used extensively in a few industries and have helped to examine and improve human reliability. However, company structures and management attitudes in many industries frustrate the gains that could be achieved in terms not only of awareness, but of reduction of accidents, both minor and serious. Perhaps the most significant of these factors is proper recognition of the importance of human reliability assurance programmes. Most managers regard accountancy, marketing and technology as subjects where the requisite expertise is acquired by systematic professional training. On the other hand, they usually believe that their common sense, instincts and practical experience will suffice in managing the human domain. Yet, analysis of major disasters shows this belief to be mistaken.

The first step towards reducing the number of major disasters (and perhaps equally important, the vast number of minor errors and their consequences which cumulatively cause more deaths and cost more money) is to develop useful concepts for the assessment and improvement of human reliability. Concepts such as Performance Shaping Factors, Mind Set, and Man-Information Interface are useful, because they indicate that human performance is capable of analysis, and that it can be shaped. What must be remembered from the outset is that the classification and analysis of such factors inevitably involves over-simplification of complex situations. While any but the most trivial classifications and definitions suffer from the same shortcoming, the problem is very significant in the human factors domain.

Awareness of the concept that human factors have a vital influence on the degree of reliability which can be achieved in a technological system leads to the recognition that the quality of engineering depends not only on technical expertise. In the selection, training and promotion of engineers additional human attributes have therefore to be taken into account. Thus, character traits and attitudes to people and to duties are criteria of particular importance for the appointment of managers.

3 CONCEPTS FOR IMPROVED ORGANISATION & MANAGEMENT (O&M) RELIABILITY

The study of Man-Machine Interfaces (MMI) has received much attention in recent years, but the concept Man-Information Interface (MII) is still awaiting wider recognition. When a person receives, transmits or interprets information, he is affected by the information. This is usually well recognised. What is not always appreciated is that the information is affected by the person. Thus, both are changed and are no longer the same as before the interface. This change in the form and content can be due to a number of transformations such as:-

    distortion
    emphasis
    addition
    deletion
    substitution
    diminution
    exaggeration

A particular message seen against a different background or perspective, or understood in a different context or frame of mind becomes a different message. It conveys a different meaning and produces a different result. Yet at each man-information interface there can nevertheless be the impression that the message has been accurately given and correctly received.

In technology management, the chain of man-information interfaces is often complex as well as difficult to control. For example, an expert system for rapid fault diagnosis may involve communication between engineers and programmers, technicians and managers. The chain of transmitted messages will involve a structure such as shown in Figure 3. The design, establishment and use of a fault diagnosis system is seen to involve many interfaces, each of which is vulnerable to transmission distortions or errors introduced by human unreliability in receiving, transmitting, interpreting and using information. All too frequently senior mangers normally assume that competent people can always be relied upon to cause no significant message degradation.

Awareness of the risk can be the first step towards guarding against the hazard. We are accustomed to think that a chain is as strong as its weakest link. This is normally true, but in a machine control chain the strength of the whole chain can be improved by applying feedback control loops which protect against the weakness of individual links. Similarly, in a management decision chain, operational audit loops and decision support systems can improve the reliability of inherently weak links, and of the chain as a whole.

4 MANAGEMENT OF RELIABILITY

There is a need to balance the economic and safety requirements, but catastrophies must be avoided with a very high degree of confidence. When a plant is not sufficiently reliable it will be unsafe or uneconomic or both.

Present techniques for reliability analysis include fault tree analysis, which is a deductive process to cover all types of potential system failure, hardware, software and human. Empirically derived data for quantitative analysis is limited, and therefore some element of judgement is still needed. Two important considerations emerge from this limitation of available reliability data:-

The first of these considerations accords with the common experience that the tasks affecting plant operation (and design) are made up of many human actions in complex, but analysable patterns. The second recognises that such tasks may occur in many parts of a fault tree, with the possibility of common influences affecting them ie dependencies between them which must be of great concern. In order to consider this further, the nature of human action needs to be carefully examined. It is also worth noting that the ultimate responsibility for organising and supervising all the tasks identified in the reliability model (eg fault tree) lies with management.

An important aspect of discharging this responsibility is the audit function, ie the systematic checking of the system and its operation. Another is to organise and manage the operation in such a way as to reduce error, eg as shown in Figure 4 where the aim is to reduce engineering error by checking functions. (The meaning of the "AND" gates is notionally the same as in fault tree terminology) eg carrying out a design review will remove many design and operational errors and defects.

Management attitudes have a vital influence on the reliability of engineering operations. Stringent completion dates, last minute changes, poor personnel relations, stress and insecurity, excessive cost-cutting and unrealistic performance specifications, all militate against the attainment of reliable engineering and manufacturing operations and against reliable products. To maintain viability and at the same time achieve reliability, a tight, disciplined and contented organisation is required, where there is mutual respect between management and staff, and which is permeated by a high standard of ethics and self-respect.

When ambitious reliability targets have to be achieved, management attitudes must be adapted accordingly and human factors will need to be taken into active consideration. Key techniques include:

- Task analysis, including task description and task representation

- Ergonomics, including classification of behavioural components and examination of failure modes

- Decision strategies based on a sound understanding of decision analysis and decision support systems

- Human error analysis

- Communications analysis

5 SOURCES OF ERROR AND CONFLICT

A general theory of the structure of action has been produced by John Searle in the 1984 Reith Lectures (8). This theory makes sense of the many issues involved especially the anomalies, and underpins some of the useful aspects of current human error models. It can be specialised so as to be useful in understanding MMI and the assessment of human error (9).

The relationship of this theory to human performance and reliability modelling has been extensively discussed in reference (9). This shows that the theory provides a firm basis for the useful models and explains many anomalies particularly relating to data. However one purpose of this paper is to show how O&M relates to human action specifically. This is through the network of intentional states described in Principle 7 of Searle's theory. According to this principle, an intentional state only 'functions' as part of a network of other intentional states. 'Functions' here means that it only determines its conditions of satisfaction relative to many other intentional states. (One doesn't have one intention in isolation). Thus, the operatives are in the control room for many reasons: personal, organisational, technical etc. The desire to successfully control a plant functions against a whole series of other intentional states eg, to maintain the reactor working, the quality of its output, please the boss, maintain the integrity of the plant, keep their jobs, job satisfaction etc. They characteristically engage in practical reasoning that lead to intentions and actual behaviour. All the intentional states together that give any specific intentional state particular meaning are called the network of intentionality.

This and the remaining principles of theory are described in some detail in references (8) and (9).

In the case of technological systems, apart from internally derived mental states eg the desire to please, satisfaction, most if not all the intentional states in the network are aspects of organisation and hence to management (responsible for setting up and running the organisation). Events, ie the internally derived states so called, relate to aspects of O&M, thus the operator function as part of a network of other intentional states involved in running the plant. These need to be properly managed for continued successful operation of the plant.

Reconciliation of conflict within the network of intentionality is of particular importance to effective risk management. The conflict between important design goals such as performance, price and reliability is a typical example, and the importance of relating supplies and user concerns is included in BS5760 Part 4 issued recently.

6 MANAGEMENT ASSESSMENT

This is the most problematic and least developed area from a risk and reliability viewpoint. It is a common influence affecting all aspects of plant operation. Some authorative sources believe that the range from very good to very poor management can produce an order of magnitude increase in risk of accidents. Some analysts believe it can best be dealt with by considering the effects of supervision, training, working environment, etc, and other management controlled factors at the detailed task level. Indeed overall control and monitoring are clearly a major management responsibility in reducing risk and improving reliability. For instance in the aviation world (10) the flight crew training programmes are expanding beyond the traditional role of maintaining piloting skills and providing instruction orientated towards flight deck management crew co-ordination, teamwork and communications.

Generally, technological management is hierarchically organised (cf management charts). So the generic problem is to relate the functions of this hierarchy to the tasks which affect plant reliability safety. A model for describing the operations of such organisational decision making has been derived from the theory of cognitive problem solving. This is a multi-level approach to the representation of decision problems. The cognitive scheme is described in detail in Reference (11) and is shown diagrammatically in Figure 5. The levels shown in Figure 5 correspond directly to the first five in Table 1 described below.

Implications of the Multi-Level Scheme for Supporting Organisational Decision Making

Table 1 shows the correspondence between the levels of abstraction involved in conceptualising decision problems described in reference (11) and Jaques' levels of abstraction of the demand characteristics of the tasks carried out by decision makers located at the various levels within the hierarchy of a bureaucratic organisation.

The qualitative differences between the levels of organisational roles shown in the first column can be understood in terms of progressive levels of abstraction in the symbolic construction of actions that may be carried out by executives at each level. Moreover, these levels are not viewed as a specific product of organisational forms; rather, bureaucratic levels are parasitic on levels of abstraction of (idealised) tasks within organisations.

The third column in Table 1 summarises the description of the demand characteristics of the tasks facing personnel with responsibility at a given level in an organisation. In any actual organisational context we may find personnel at particular organisational levels also responsible for carrying out tasks at lower levels (rather than delegating them to subordinates). Executives may be able to take initiatives at more than one level in organisations where the role structure permits this (for example, as "consultants" or "problem fixers").

In a practical analysis the scheme in Table 1 will also be divided horizontally into columns representing different management activities or agencies such as resource management, safety assurance, QA, personnel operations etc. The communications between these vertical "lines" of management as well as between levels will be of crucial importance to the success of plant operation. This theoretical framework is now being developed in a study of the management of inspection processes and utilised in the assessment of the management of industrial systems.

208

7 <u>REVIEW OF MAJOR ACCIDENTS</u>

Case studies of recent catastrophic accidents demonstrate the close connection between management organisational shortcomings and the causation of the disaster.

Reviews of three such major accident reports show many similar features which may be summarised from an O&M viewpoint as follows:-

a. Challenger (3)

   1. Lack of clear corporate safety organisation.

   2. Lack of adequate communications down the line and between lines eg between project managers and flight operations.

   3. Increased flight rate.

   4. Decreasing resources.

   5. Design control issues.

   6. Maintenance management problems.

b. Bhopal (4)

   1. Lack of a clear safety assurance policy.

   2. Lack of operator knowledge ie training.

   3. Decreasing resources.

   4. Design support problems.

   5. Maintenance management problems.

c. Chernobyl (7)

   1. Problems in the management of the safety bureaucracy related to communication difficulties between higher and lower levels and in the allocation of responsibilities.

   2. The pressure of demand for electricity was very high and Chernobyl was at the top of the "plant availability" league.

   3. Acknowledged lack of operator training.

   4. Extreme pressure of time.

   5. Design shortcomings.

This listing indicates the lack of appreciation by management and engineers involved of the significance of these issues. These critical O&M issues have been highlighted in the last column of Table 1. This also indicates at what levels responsibility for their issues might reasonably be expected to be substantial. Clearly this may vary from organisation to organisation. Assessment however should demonstrate whether appropriate functions are being performed and whether there are any weaknesses in functional performances or in communications between and across levels.

8   METHODS OF ASSESSMENT/ANALYSIS

The framework derived from the hierarchical cognitive theory described and the evidence of accidents may be put to use in two ways:-

1.   As part of safety/reliability assessments of systems to provide an appraisal of the organisation and management of that system.

2.   As an aid to accident analysis which structures the information and evidence arising from an enquiry.

These two uses are briefly described below:-

Assessment Method - O&M Appraisal

For this to be most useful it should be part of a safety/reliability assessment, so that detailed technical evidence of possible failures of the system are available some of which will impinge on or be caused by O&M. The overall objectives of the system should be clearly defined. Hierarchical organisation charts should be available together with descriptions of the responsibilities and tasks of the staff involved at all levels of the O&M tree. The O&M structure is then matched to the framework given in Table 1 column 2 to get the best fit possible. Precise fits are not always possible, but generally one or two levels of management will fit to each level in the table, or to a merger of two levels. The responsibilities and tasks at each level in the matched structure are then examined against the criteria in the third, fourth and particularly the fifth column where each of the items eg maintenance should be considered separately in turn. This is done against a background of information arising from the safety/reliability analysis eg fault trees, event trees etc, which will show where in the system operation many important failures including human ones could occur. The assessor then has to judge whether the maintenance etc, O&M is adequate at each level, eg has the design been carried out by a competent organisation providing adequate records, product assurance covers QA and reliability, is safety adequately monitored. There will be differing levels of responsibility for each item at each O&M level and this will vary between organisations and systems. In the end the assessor uses the framework for:-

1.   assessing the adequacy of the O&M.

2.   to identify any 'holes' in the O&M which could be important to S&R and to agree with management on recommendations for dealing with problems.

Clearly this method is qualitative, but it provides a background against which confidence in the quantitative S&R assessment of the system/plant may be guided. It could affect the choice of reliability data, the assessment of operator error, the assessment of system dependencies etc. It may be useful to utilise task analysis methods particularly at the lower

210

levels of O&M where they are more applicable in order to clarify the operation of the plant. This may also arise from the S&R analysis. At higher levels the use of the Management Oversight Risk Tree (MORT) method may be applicable and this is being considered for further development of the method.

Providing a case study or example of this method is difficult because of the confidential nature of the information involved, however, the issues may be better understood by considering the second use of the framework and taking the example of a well known and publicised major accident, the Herald of Free Enterprise Ro-Ro ferry disaster at Zebrugge in 1987.

Accident Analysis

The important aspects of the Ro-Ro ferry disaster are publicly known and these will be entered into the framework in the following brief analysis.

At level 1 there was the bosun who was asleep (loss of resource) and clearly failed in his <u>safety duties</u> concerning the closure of the doors. The second officer at level 2 did not check properly that the bosun was present (safety and resource problems). They both incidentally should be aware of the operating states of the doors (maintenance function). At level 3 the captain was not provided with adequate means to check door closure (design) nor was there an adequate system for doing this ie (PA) (product assurance). Although he was aware of the safety aspects, the safety procedures for which he must be partly responsible at least were poorly defined. The ship's crew resources were also inadequate to ensure that the doors were closed when they should be, because of conflicting duties, so there was a failing at this level of O&M and above as it turns out. With regard to structuring capability, clearly the risks being taken at each level were not understood, not reported to the next level, there was lack of sensitivity to the relevant issues and in fact this appeared to go right up to level 6/7.

Level 4 general management failed to provide clear safety directions and ignored protests by some captains on a number of points, eg plimsoll line checks, number of passengers, door closures. There was lack of PA, eg passenger checks, deliberate overriding of sound safety recommendations, eg door closure warning lights and enforcement of mandatory procedures. At level 5 there was a lack of a clear safety policy, the risks associated with the ship design in operations were not articulated and the relevant principles of operation spelt out, so adequate resources were not provided and no adequate PA system defined. At levels 6 and 7 the implications of the ship design had not been fully appreciated from an operational view-point and overall strategy for the fleet safety, PA and resourcing promulgated. There may have been confusion on understanding of corporate and ships' masters responsibilities which affected levels 5, 4 and 3.

9   CONCLUSIONS

The accidents considered have similar generic features, eg resource problems, safety management inadequacies, communication problems, design issues, and inadequate information. These could potentially have been assessed and led to accident prevention by the methods discussed in the paper. The O&M assessment would have enabled common features to be identified that could have affected many aspects of plant operation, eg demand pressures/resources, possible deterioration of maintenance standards, limitations of design, to be considered in the task analyses and the effects to be allowed for in the safety/reliability analysis and assessment. It is clear that major accidents cannot be accounted for on the basis of a snapshot approach. Variation over a long period of time has to be considered. This means continual review of safety and reliability assessment. It also means that reliability data based on certain standards of plant maintenance for instance are not applicable in potential accident occurrences. Thus not only are there increasing dependencies but item failure rates are likely to be increased also.

These analyses of O&M influences on plant reliability will be mainly qualitative. This is because of their mainly mental and social characteristics which make quantitative data collection and analysis fundamentally difficult. This also applies to the reliability of management itself, eg in the communication process or in controlling resources properly that may affect plant reliability. The issue of avoiding catastrophic accidents is too important to thereby underestimate the importance of performing such analyses since they can point to ways of preventing common influences which give rise to unforeseen dependencies.

Catastrophic accidents are analysed publicly in depth much more extensively than commercial disasters (perhaps Nimrod will be exceptional) in less dramatic situations, and from such analyses factors emerge that have application experience when considered objectively also help to identify areas where problems arise. It is acknowledged generally that design should be properly controlled to ensure reliability and that users should follow recommended procedures if they are to avoid losses. Indeed at the heart of BS5760 on Systems Reliability Management, Part 4 on the specification of reliability is the concept that designers and users should clearly understand the life cycle requirements of equipment and act accordingly.

A wide gap exists at present between specialists and managers. The human factors specialist and his expertise are remote from senior management and the topics of concern to him are isolated from the practice of high level decision making. In consequence, few opportunities exist for putting new knowledge in human reliability into the hands of those who can apply with the greatest advantage. Moreover, because of this gap, research workers lack adequate, direct input from the potential user on senior management level who alone can help to focus effort on the most important issues, and provide the necessary feedback from practical application. Progress in this field is therefore slower and knowledge gained less relevant than it would be if this gap did not exist.

It is with these considerations in mind, that the Human Reliability Factors in Technology Management Project Group was formed by the Safety and Reliability Directorate, with the hope that its work will go some way towards building a much needed and effective bridge. The most important aim of the group is to initiate, foster and participate in projects to increase relevant knowledge of Human Reliability Factors and apply these in the practice of technology management. The group is also intended to bring together participants from a wide spectrum of experience, ranging from specialised research to general management in industry, and from junior executive to senior board level.

The subject matter of concern to the Project Group is focused on the management of technology where significant risk is involved with respect to safety or commercial viability, and where reliability is important. The aspects of management of concern are closely related to the design, construction and operation of industrial plant and products.

A number of projects are in the process of definition and initiation. A management risk guide is the first priority. Other projects include decision support systems for safety and reliability management, assessment of the management contribution to risk taking, hierarchical modelling of management structures affecting plant reliability and analysis of attitudes towards risk. Anyone professionally concerned in this area and interested in the group should contact the authors.

References

1   Contents of the NCSR Reliability Data Bank NCSR/DB/40.

2   Watson, I A. "Review of Human Factors in Reliability and Risk Assessment" I Chem E SYMPOSIUM Series No 93.

3   Covallt Craig. "SHUTTLE 51-L Loss". Aviation Week and Space Technology/June 16, 1986.

4   Bellamy, L J. "The Safety Management Factor: An Analysis of the Human Error Aspects of the Bhopal Disaster".

5   Edwards, G T et al. "Defences Against Common-Mode Failures in Redundancy Systems". SRD R196, January 1981.

6   "Human Factors in Industrial Systems - Review of Reliability Analysis Techniques". (Draft with the HSE, Bootle, Liverpool).

7   Collier, John G et al. "Chernobyl" CEGB September 1986.

8   Searle, John. "Minds, Brains & Science". The 1984 Reith Lectures. Published by the British Broadcasting Corporation.

9   Watson, I A. "Fundamental Constraints on some Event Data" Proceedings of the 5th EuReDatA Conference Heidelberg, April 1986.

10   Aviation Week & Space Technology. October 1, (1984) Page 99. "Cockpit Crew Curriculums Emphasise Human Factors".

11   Humphreys, P and Berkeley D. "Handling Uncertainty Levels of Analysis of Decision Problems" in G Wright (Ed) "Behavioural Decision Making" London: Plenum Press 1985.

LEVEL	MANAGEMENT ROLE	JOB REQUIREMENT	STRUCTURING CAPABILITY	RESOURCES	PRODUCT ASSURANCE	MAINTENANCE	DESIGN	SAFETY
7	CORPORATE MD	LEADING CHANGES IN POL/SOCIO/TECH/STRATEGY	ASSESSMENT OF ANY CHANGES IN LEVEL 5	X				X
6	CORPORATE EXECUTIVE	CO-ORD & TRANSLATION OF CORPORATE STRATEGY	ANALYSIS OF SPECIFIC OPERATION IN 5	X	X	X		X
5	DIRECTOR OF ENTERPRISE	DETAILS AND MODIFICATION OF 6	ARITCULATION OF ORGANI-SATIONAL PRINCIPLES & OVERSIGHT OF IMPLEMENT $\underline{N}$	X	X	X	X	X
4	GENERAL MANAGEMENT	IMPLEMENTATION AND DEV OF BUSINESS	SELECTION & INTERFACING OF DIFFERENT ORGANISATIONS	X	X	X	X	X
3	LOCAL MANAGEMENT	CONTROL OF TRENDS FORMULATION OF PROBLEMS	RESTRUCTURING OF FUNCTIONS		X	X	X	X
2	FRONT LINE MANAGEMENT	ANTICIPATE CHANGES AND LOCAL CONTROL	SENSITIVITY ANALYSIS ON LEVEL 1			X		X
1	OPERATIONS	CONCRETE TASK AT HAND	ASSESS/REPORTING WITHIN FIXED SYSTEM			X		X

MANAGEMENT RESPONSIBILITIES & STRUCTURING CAPABILITY

TABLE 1

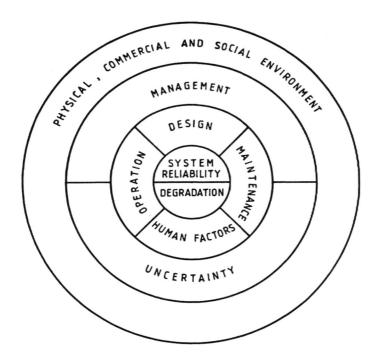

FIG. 1 FUNDAMENTAL FACTORS IN SYSTEM RELIABILITY

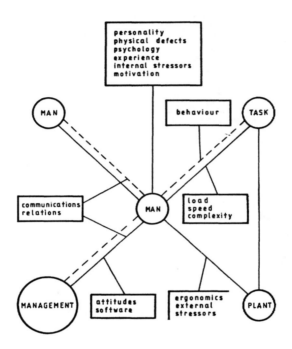

FIG. 2   INFLUENCES ON MAN IN INDUSTRY

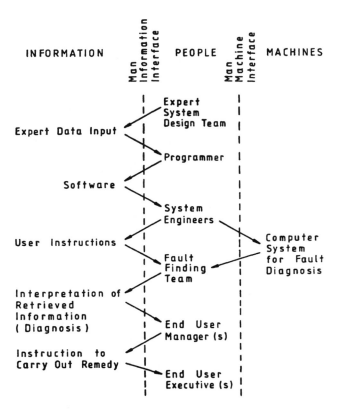

Each arrow in the chain signifies a crossing of the interface at which changes and errors may be introduced inadvertantly.

FIG. 3   SYSTEM INVOLVING MAN-INFORMATION INTERFACES (MII)

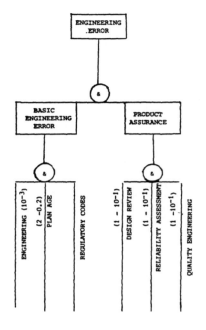

FIGURE 4   REDUCTION OF ERRORS

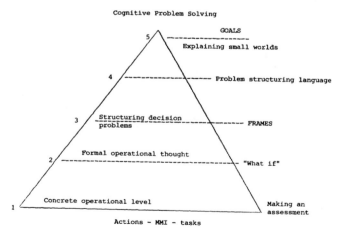

Fig 5

217

PROCESSOR-BASED DISPLAYS: THE FLEXIBLE CONTROL PANEL

T. F. MAYFIELD
Electrical and Electronics Design Department,
Rolls-Royce and Associates Limited,
PO Box 31, Derby, DE2 8BJ

ABSTRACT

Hard-wired Nuclear Power Plant (NPP) panels, however ergonomically
designed, are comparatively inflexible across the whole range of operating
conditions. Thus panels will, in general, have been designed for the
operating modes most usually adopted. There will only be room, within the
operator's immediate arc of vision and control, for a limited inventory of
controls and displays, for those operating modes most usually undertaken.
The correct use of computer-based displays, however, means that the
operator can be given the right operating information, even for rare
events, on demand and in the main display area. Whilst this may be
expected to improve operator reliability, guidelines on how such displays
should be presented are either not available or are too general for
specific applications. This paper discusses the development of dynamic VDU
displays for NPP.

INTRODUCTION

Conventional Panels

Prior to Three Mile Island (TMI) NPP Control Rooms generally consisted of
large panel suites with many hundreds of Control and Instrumentation (C&I)
channels. Such hard-wired control systems were directly linked to the
parameter being measured or the equipment under control. System
reliability was dependent on the fact that the instrumentation channels
were one to one (detector to display) and, on safety related systems were
duplicated and/or part of a majority voting logic; this gave a high degree
of operator confidence. High priority alarms were also on a one to one
basis, being important in the initiation of emergency routines. The only
exception to this was for non-critical, warning channels, which were
usually fed via some form of multiplexing system; each channel being
accessed, in turn over a very short period of time.

218

The accident at TMI, with its emphasis on operator error; the ever increasing scale and complexity of NPP systems, requirements for operator training and changes to manning levels has meant that attention to the operator interface has become essential. Traditional ergonomics provides plenty of guidelines for the design of control panels (Refs 1 to 2) with much of the work to date centering around the following areas:

a) Specification of physical and environmental aspects; for example, anthropometrics, vision, lighting, noise and ventilation.

b) Optimisation of control and display integration; for example, illuminated mimics incorporating C&I items.

c) New display devices; for example, processor driven displays.

The careful application of sound ergonomic principles to the design of panels, can help avoid some of the problems of human reliability, and post-TMI work in the USA has concentrated on such areas. However, it is very difficult to both represent every parameter on a one-to-one basis and satisfy good ergonomic design at the same time. The sheer scale of the problem means that any layouts are a compromise when considering the possible range of operating conditions. Consequently, panels tend to be well designed for normal operating conditions (eg load operations) but may not cater effectively for shutdown, emergency or abnormal operations. Due to this inflexibility, operators will be at a disadvantage when it comes to reacting to emergency or unfamiliar operating modes. Experience shows that real-life incidents are often very different from the perceived emergencies practised in the classroom or on simulators.

The concern felt, both inside and outside the nuclear industry, about the control of such complex plants, and the problems of quantifying human reliability, have made it inevitable that more and more automation and computerisation would take place. Indeed, the development of Safety Parameter Display Systems (SPDS) has progressed from hardwired, through multiplexed to processor-based systems. However, there is a danger that the greater availability of information will lead to further overloading of the operator. The work described in this paper is from a Study addressing these issues.

Software Based Displays

The greater power and flexibility of the latest software-based displays can, if designed with care, decrease the load on operators. It is now possible to progress (with caution) to the point at which the operator can be given only that information which he needs, when he needs it and, more importantly, where he needs it.

The use of computers, and their more obvious "face" the Visual Display Unit (VDU), is quite common in process control, especially for alarm and warning sytems. However, this has often meant converting channel analogue signals to digital, with a consequent rise in panel components. The development of serial highways and digital instrumentation makes the use of processorbased displays more attractive.

Such developments need to be introduced slowly and with due consideration for risk assessment and safety justification. Typically there has been a piecemeal introduction whereby, initially, only display information is software based. This is then followed by converting to software some control functions that are non-safety critical, the major requirement being to ensure that the reliability of the software based systems, for safety related parameters, is to the same standard as previous hard-wired requirement specifications.

STUDY AREAS

Study Aims

Unfortunately, whilst the enabling technology is available, the amount of ergonomics guidance on the formatting of such displays is limited (Refs 3 and 4). This is particularly true where dynamic, pictorial representations are used. Traditional engineering standards, such as those for Piping and Instrumentation (P&I) drawings, do not always convert easily to the VDU screen. In addition, there is little empirical data on whether operators find the current VDU display designs more or less easy to use than the current hard-wired panels.

General human factors standards and guidelines can optimise interface design. However, the specific context (type of plant, operational constraints etc) and the particular tasks being undertaken (start up, shut down, emergency operations etc) will modify (and, even, conflict with) any general considerations. This study is part of a long term investigation into software generated displays. The aims of the study are:

a) To develop the concept of, and produce, scenario-based displays, that is displays which reflect the operation under surveillance. This enables the operator to have, in his prime viewing area, all the information relating to the current operation.

b) To derive display formatting rules for the development of dynamic, graphical displays. The displays would be of both the scenario and systems mimic type.

Scenario-Based Displays

To develop the display formats a task and information analysis was used and was itself instrumental in defining the concept of the scenario based display. Although only a limited number of procedures could be examined, they were felt to be representative of the main activities carried out by the operator.

Four procedures, involving a high level of operator interaction with the plant were selected; providing a wide range of operator interactions and covering normal and emergency operations. These specific procedures were selected because they covered many of the plant and equipment line-ups likely to be experienced during the plant operating life.

A step-by-step analysis was carried out on each procedure in order to determine the control and instrumentation requirements and the task sequences. Operator experience was used to evaluate the time and sequence relationships, governing each operator's actions. A number of "common sequence" or "procedural loops" were identified and found to occur at regular intervals. These common sequences often crossed system boundaries in terms of the components represented.

Four major "procedural loops" were identified as follows:

a)   Rod Control.

b)   Primary Coolant Make-up and Discharge.

c)   Electrical Synchronisation.

d)   Feed Water Control.

Although these are the key sequences found, the restricted nature of the analysis so far, means that there are probably other procedural loops which remain to be identified.

From the basic requirements of these sequences, formats for the display pages (including the relevant controls) can be derived. These pages will represent the major operating scenarios and provide all the necessary information for the control of the plant.

Display Formatting

Before finalising any of the scenario displays some work needed to be done on the rules for formatting. Basic information on text size, brightness, colour etc can be found in a number of texts, for instance, the VDT manual (Ref 5). However guidelines on dynamic displays using active mimic diagrams, real time plant variables, trend and history graphs etc are not yet available in a format which can be used by system designers. This part of the study is being carried out to try and define rules for the following areas.

Flowlines: Representation of pipelines and electric circuits. Key issues are width of lines; connectivity coding (via valves, breakers etc); availability and status.

Valves: Key issues are coding (i.e. colour, symbology, orientation etc.) for open, shut, tripped; connectivity; availability.

Pumps: Key issues are coding for on, off, standby, tripped; flow direction and magnitude; connectivity, availability.

Vessels: Representation of tanks, steam generators, condensors, flasks etc. Key issues are coding for state; connectivity; availability.

Machines: Representation of generators, motors, air conditioning equipment etc. Key issues are coding for status, connectivity, availability.

Screen Formats: Key issues are density of characters and graphics components; layout of displays; use of menus, windows, icons; coding methods – colour, size, shape; use of mimics.

Alarms and Warnings: Key issues are priority; status; availability.

Access: Key issues are levels; navigation; number of interactions; manual and auto demand.

DISPLAY TRIALS

To test out the usability of the display formatting rules, a series of trials is being undertaken. Plant operators will be asked to evaluate various component formats for ease of recognition, legibility, applicability and so on. Criteria for evaluation will be quickness of response, accuracy of identification and subjective like or dislike. Trial displays are currently being formatted to test the following:

a)   Different methods of indicating open and shut valves.

b)   Different methods of indicating pumps on and off.

c)   Different mimic line thickness.

d)   Use of black versus grey page background (other background colours may also be tested).

e)   Different methods of displaying analogue data.

f)   Different methods of displaying alarm and warning information.

It is hoped that the results from these trials will provide the basis for the display format rules to be applied to the scenario based pages.

CONCLUSIONS

This study is still in its early stages but progress to date is encouraging. The value in organising displays around the current operating scenario is already evident. Operators will not need to be tied to inflexible panels which only supply all the necessary information for a small percentage of the time. However, much work still remains to be done in validating the concept. In addition, the arrangement and numbers of displays and the lack of an overall indication of plant state need further consideration.

Valuable experience in the design of components for VDU pages is being gained and the problems inherent in some display designs (clutter, lack of logical grouping, poor coding, poor use of colour etc) will be avoided.

The current situation can be likened to the state of affairs in panel design prior to the widespread application of human factors. Then it was a reluctance on behalf of engineering designers to take note of the available techniques and, on behalf of the design authorities, to initiate fundamental research to relate the general human factors principles to the specific context of panel design. The accident at Three Mile Island, Unit 2, in the nuclear industry, and other major accidents in the non-nuclear industry, for example Bhopal, where human error was considered the prime cause, has prompted a high level of investigation into incorporating ergonomics at the design stage.

It is against this background that the current work is being undertaken, so that the undoubted advantages of technical advances in C&I techniques are not lost due to poor or unusable interfaces.

REFERENCES

1. McCormick, E.J. and Sanders, M.S., _Human Factors in Engineering and Design_, McGraw-Hill, New York, 1982.

2. _HF Guidelines for NPP Control Room Development_, EPRI NP-3659 Essex Corporation, Palo Alto, CA, 1984.

3. Martin, J., _Design of Man-Computer Dialogues_, Prentice-Hall, New Jersey, 1973.

4. Shneiderman, B., _Designing the User Interface_, Addison Wesley, Reading, Mass, 1987.

5. Cakir, J., Hart, D.J., Stewart, T.F.M, _VDT Manual_, 1984.

223

A HUMAN FACTORS DATA-BASE TO INFLUENCE
SAFETY AND RELIABILITY

JEREMY C WILLIAMS
Sizewell 'B' Project Management Team
Central Electricity Generating Board
Booths Hall
Knutsford
Cheshire
WA16 8QG

ABSTRACT

Throughout a project's life the success with which a product or service is delivered is strongly influenced by human beings. Their performance is determined by the nature of the tasks they are called upon to perform and the circumstances under which such demands are made.

It is with this knowledge in mind that, for the last decade, Probabilistic Safety and Reliability Assessors have had a pressing requirement for a viable human reliability data-base from which they can make plausible deductions.

This paper considers the question of why it is that attempts to collate human reliability data have tended to fail so far, and what it is that can be done to build a viable data-base in the immediate future. The paper also demonstrates what is currently known and describes the basic work that appears to prove the principles for further development of a human reliability data-base, which, when applied, may be expected to influence operational decisions and the assessment of resistance to failure.

INTRODUCTION

Safety and reliability engineers, whilst recognising the need for a human factors data-base have, nevertheless, discovered that such information as they can find, tends either to be over-simplified or extremely complicated. The simplification of such 'data' can be viewed as being a mixed blessing. Although the concepts seem easily expressed there is a concern that the real data underpinning such broad assertions may not support the generalisations. The complicated view of human

reliability 'data', on the other hand, suggests that the probabilities
of failure are so difficult to assess because of the complexity of
performance shaping factor interactions, that no worthwhile judgements
can be made.

Whilst both these views are understandable, and perhaps held with
equal conviction, they do not assist the development of a viable human
factors data-base for the use of safety and reliability engineers. Both
views encourage us to look no further. The over-simplified view creates
the impression that there is no need to investigate the requirements
for a human factors data-base because the information that is available
is adequate. By way of contrast the complicated view discourages
exploration because it presupposes that the subject is so daunting that
no progress could be anticipated, even if effort were applied.

This paper accepts neither extreme view, but suggests that there
is a need to continue efforts to develop a human factors data-base even
if the two views, for quite different reasons, argue otherwise.

There can be little doubt that those arguing from the simplistic
viewpoint would be interested to learn more about our real knowledge
in the data area, even if they were to decide, subsequently, to ignore
it. Likewise those arguing from a complexity viewpoint would also be
interested, if only to confirm their view that the problem is almost
intractable.

The author believes that progress in absolute data collection and
organisation is likely to remain a central problem for the next ten to
fifteen years, but submits that sufficient is now known about relative
data in the human factors domain, that some attempt to organise it is
justified.

This belief is supported by the often-expressed view that although
many assessors would like to have fully quantified data, most would settle
for any credible information, even if it were only at a qualitative
level.

The author contends that the human factors literature, which now
extends back in time some forty years or so, possesses many of the answers
which safety and reliability assessors seek. Although data collected
during the countless studies reported exhibit massive absolute
variability, they possess very much narrower relative variability. So
narrow is this relative variability that it is possible (to a first
approximation) to describe major performance modifying effects in very
simple, consistent mathematical terms, so that reliable predictions can
be made about the probable extent of an error probability change, given
that the situation can be modelled with some clarity.

It is this portion of the existing human factors data-base upon
which most of this paper is based. The construction of an absolute human
reliability data-base looks feasible but is likely to require much more
information to develop than currently appears to be in the public domain.
As relative data are much more abundant, however, it is the creation
of a relative data-base to which much of this paper is devoted.

HUMAN RELIABILITY DATA-BASES

Quite a number of attempts have been made to collate human reliability data. A good many of these have concerned themselves with absolute probabilities and appear somewhat implausible for a variety of reasons.

Three basic types may be discerned from the literature. One class relies upon expert judgement to interpret somewhat vague basic data that are not always supported by traceable references. A second type has strong relationships to experimental or actuarial data and has rather specific characteristics. The third set attempts to use expert judgement calibrated in some way to extend or complete knowledge in particular areas of concern.

The first type of data-base [1, 2] can sometimes appear to be too general to have much applicability, and rarely has much explicit support. The second type of data-base [3, 4] often has highly specific characteristics which are difficult to generalise, and the third type [5, 6] frequently appears to offer promise but only in the specific context of the extrapolation, interpolation or calibration applied by its authors at the time of its creation.

Thus we seem to have considerable difficulties. Most of the human reliability data-bases available at present appear to have many of the deficiencies of the worst kinds of data-base, with few of the benefits of the better kind. From the evidence available at present it appears that in some respects the wrong problem may have been tackled.

The shortcomings are fairly easy to articulate, but what is more difficult is finding something to put in front of Probabilistic Safety and Reliability Assessors that has experimental or actuarial data support, some element of generalisability, but sufficient specificity that they can recognise when and where the information would be applicable.

Although the need is ultimately for an absolute probabilistic data-base it seems obvious that, for logistic reasons in the immediate future, progress towards this objective will be slow. Therefore, this author has concentrated on a human factors data-base to influence safety and reliability which is responsive to the concerns of assessors, has some experimental or actuarial support and elements of generalisability and specificity with which they can identify.

Relative Data – The State of Knowledge

Although originally presented as an absolute data-base, the most obvious example of a relative human reliability data-base is the American Institutes for Research (AIR) Data Store [3]. To create this data-base its creators searched some 2000 references, deciding eventually to use 164 as inputs to their judgements regarding the absolute probability of success when using a variety of controls and displays. The relative differences that were predicted are implied in the probabilities quoted in the AIR Data Store.

Other examples are the data-base known as the Bunker-Ramo Tables [7] (said not to reflect performance shaping factors (though in many ways they do)), the 'data' generated by Pontecorvo [8] using a similar methodology to that employed by Irwin et al [7], the graphs suggested by Bello and Colombari [9] and the likelihood of accomplishment scale [10] which also attempted to give insight into the relative effects of various factors on human performance.

Most previous attempts to present human factors relative data may be characterised by their specificity to man-machine interface details, their rather 'expert' origin or their somewhat limited scope.

This paper attempts to extend the range of such data in all respects and gives indications that with sufficient perseverance it should be possible to build quite a wide-ranging and robust relative human factors data-base for relatively little outlay of effort.

Much of the thinking is analogous to that used by the author to develop the Human Error Assessment and Reduction Technique (HEART) methodology [11]. A concerted effort has been made to examine the significance of a range of human factors experimental literature as it relates to the apparent needs of the safety and reliability community. Their desire to understand the extent to which human factors phenomena might affect decision-making with respect to operations and the assessment of systems design resistance to failure has been recognised, and where possible, a statement as to the likely nominal strength of the phenomenon has been made.

A wide range of human factors literature has been examined with a view to extracting the salient features relevant to safety and reliability engineers. Patterns are apparent in this literature. Even though in an absolute sense the data seem quite variable, in a relative sense the changes to human reliability which various factors produce are remarkably consistent. The approximate strengths of these factors are indicated in an aggregate sense.

The factors are described using the same terminology as has been used in the HEART methodology and estimates regarding their industrial significance are given. There are some similarities between the method adopted and that used by Teichner and Olson [12], although it should be noted that few data were available at the time of their development work. Teichner and Olson's interest was in absolute rather than relative data, and it is perhaps significant that no further successful work has been undertaken with regard to absolute data aggregation since Teichner and Olson's original attempts. This is thought to be because of the wide variability of such data. Relative data however, exhibit remarkable within-factor consistency, and it seems likely that this set could be the first of many attempts to set down what is known at a factorial level.

Relative Human Reliability Data –
What The Human Factors Literature Tells Us
The following data set represents a first attempt to piece together a giant jigsaw. It is recognised that many iterations will be required before a completely appropriate data-base can be constructed. In the meantime it is hoped that this data set will serve as a catalyst to others to challenge the deductions of the author, marshall further information to fill in the blanks, extend the argument and propose alternative

strategies to set down in usable form, some of what the human factors community has learned from 40 years of research.

Factor influencing Human Reliability (Error-Producing Condition)	Maximum nominal amount by which unreliability appears to change	Industrial Significance
A. Unfamiliarity with a situation which is potentially important but which only occurs infrequently or which is novel [13, 14, 15, 16]	x 17	v great
B. A shortage of time available for error detection and correction [17, 18, 19, 20, 21, 22, 23]	x 11	great
C. A low signal-noise ratio (when really bad) [24, 25, 26, 27, 28]	x 10	great
D. No means of conveying spatial and functional information to operators in a form which they can readily assimilate [29, 30, 31, 32, 33, 34, 35, 36, 37]	x 8	strong
E. A mismatch between an operator's model of the world and that imagined by a designer [38, 39, 40, 41, 42, 43, 44, 45]	x 8	strong
F. A channel capacity overload, particularly one caused by simultaneous presentation of non-redundant information [46, 47, 48, 49, 50, 51, 52, 53]	x 6	strong
G. A need to unlearn a technique and apply one which requires the application of an opposing philosophy [54, 55, 56, 57]	x 6	strong
H. A mismatch between perceived and real risk [58, 59, 60, 61, 62, 63, 64, 65, 66, 67]	x 4	measurable
I. Poor, ambiguous or ill-matched system feedback [68, 69, 70, 71, 72, 73]	x 4	measurable

J.	No clear, direct or timely confirmation of an intended action from the portion of the system over which control is to be exerted [74, 75, 76, 77]	x 4	measurable
K.	Operator inexperience (e.g. a newly-qualified tradesman, but not an 'expert') [14, 15, 45, 78, 79, 80, 81, 82, 83]	x 3	measurable
L.	An impoverished quality of information conveyed by procedures and person/person interaction [84, 85, 86, 87, 88, 89, 90]	x 3	measurable
M.	No diversity of information input for veracity checks [27, 91, 92, 93]	x 2.5	measurable
N.	A danger that finite physical capabilities will be exceeded [94, 95, 96, 97, 98]	x 1.4	comparatively small
O.	Evidence of ill-health amongst operatives, especially fever [99, 100]	x 1.2	comparatively small
P.	A poor or hostile environment (below 75% of health or life-threatening severity) [101, 102, 103, 104, 105, 106, 107, 108, 109]	x 1.15	comparatively small
Q.	Prolonged inactivity or highly repetitious cycling of low mental workload tasks [17, 99, 110, 111, 112, 113, 114, 115, 116, 117, 118, 119]	x 1.1 for 1st half-hour x 1.05 for each hour thereafter	comparatively small
R.	Disruption of normal work-sleep cycles [120, 121, 122, 123, 124, 125, 126, 127]	x 1.1	comparatively small
S.	Task pacing caused by the intervention of others [128, 129, 130, 131, 132]	x 1.06	comparatively small
T.	Additional team members over and above those necessary to perform task normally and satisfactorily [133, 134]	x 1.03 per additional man	comparatively small

DISCUSSION

The data-base which has been presented is fairly robust. The literature appears to confirm that human unreliabilities may be expected to change by the nominal amounts indicated, although of course from situation to situation the precise magnitude of the change will depend upon the proportion of the error-producing condition applicable at the time.

Previous attempts to collate human reliability data appear to have failed because they have sought to aggregate highly dissimilar absolute probabilities. It is clear from the human factors literature that such convergence, if it occurs, is not yet supported by published findings, except in a general sense.

A viable data-base, however, can be built now, if it utilises relative rather than absolute data. The consistency of the relative effects exhibited by the literature is quite strong, and judged by the author, strong enough to merit attempts to aggregate the effects into some coherent form.

This paper has shown that it is possible to amalgamate relative data from a wide range of human factors sources with a reasonable degree of success. It would not be difficult to apply this approach to a variety of human performance influencing factors. For example, factors such as the arrival rate of information, the potential for accident repetition, aging, time of day, 'user friendliness', information recognition and recall, motivation, depth of computer hierarchies, error detection, duration of information display and 'visual noise' all appear to have a rich literature which could be tapped without massive additional effort to extract salient details and the strengths of relevant performance influencing factors.

CONCLUSIONS

It has been argued that although safety and reliability engineers would like to have an absolute human reliability data-base, most would probably settle for a relative data-base, especially if this can be under-pinned by substantial evidence.

An attempt has been made to aggregate relative information from the human factors literature to portray the significance of a range of performance modifying effects. It is considered that this aggregation demonstrates remarkable consistency within and between factors, and that this consistency is sufficient to justify further attempts to consolidate our knowledge of the human factors literature in a systematic fashion. A fully developed human factors data-base of the sort described would be sufficiently robust in a relative sense that it may be expected to have profound importance to safety and reliability engineers in the ways in which they can influence operational decisions and improve the resistance of systems to failure.

ACKNOWLEDGEMENTS

The author would like to thank Miss J L Askey and Mrs M E Brotherton for preparing this paper, and Mr B V George, Project Technical and Director of the Sizewell 'B' Project Management Team, CEGB, for permission to publish it. The views expressed are those of the author alone.

REFERENCES

1. Kletz, T.A., Human Error - some estimates of its frequency. Safety Note 74/7, paper presented at a Symposium on Reliability Engineering for Instrument Engineers, Welwyn, 15 May 1974, Imperial Chemical Industries Limited, Billingham.

2. Hunns, D.M. and Daniels, B.K., Paired comparisons and estimates of failure likelihood. Design Studies, 1981, 2, 9 - 18.

3. Joos, D., Sabri, Z. and Husseiny, A., Analysis of Gross Error Rates in Operation of Commercial Nuclear Power Stations. Nuclear Engineering and Design, 1979, 52, 265 - 300.

4. Payne, D. and Altman, J.W., An Index of Electrical Equipment Operability: Report of Development, American Institutes for Research, January 1962.

5. Swain, A.D. and Guttmann, H.E., Handbook of Human Reliability Analysis with emphasis on Nuclear Power Plant Applications, NUREG/CR-1278, Sandia Laboratories, US Nuclear Regulatory Commission, 1983.

6. Ablitt, J.F., A Quantitative Approach to the Evaluation of the Safety Function of Operators in Nuclear Reactors, Report AHSB (5) R160, Health and Safety Branch, UKAEA, 1969.

7. Irwin, I.A., Levitz, J.J. and Freed, A.M., Human Reliability in the Performance of Maintenance, Bunker-Ramo Corporation, Report 317-BSD-TDR-63-218, 1963, and Report LRP 317/TDR-63-218, Sacramento, California, Aerojet-General Corporation, May 1964.

8. Pontecorvo, A.B., A Method for Predicting Human Reliability, Annals of Reliability and Maintainability, 1965, 4, 337 - 342.

9. Bello, G.C. and Colombari, V., The human factors in risk analysis of process plants: The control room operator model "TESEO", Reliability Engineering, 1980, 1, 3 - 14.

10. Hornyak, S.J., Effectiveness of Display Subsystem Measurement and Prediction Techniques, Griffiss AFB, New York, Rome Air Development Center, Report TR-67-292, October 1967.

11. Williams, J.C., HEART - A proposed Method for Assessing and Reducing Human Error. In *Proceedings of the* 9th *Advances in Reliability Technology Symposium,* University of Bradford, 1986, pp. B3/R/1 - B3/R/13.

12. Teichner, W.H. and Olson, D.E., A preliminary theory of the effects of task and environmental factors on human performance, *Human Factors,* 1971, 13 (4), 295 - 244.

13. Hawkins, J.S., Reising, J.M. and Woodson, B.K., A Study of Programmable Switch Symbology. In *Proceedings of the Human Factors Society - 28th Annual Meeting,* 1984, pp. 118 - 122.

14. Barfield, W., Expert-novice Differences for Software Implications for Problem-Solving and Knowledge acquisition, *Behaviour and Information Technology,* 1986, 5/1, 15 - 29.

15. Williams, J.C. and Willey, J., Quantification of Human error in Maintenance for Process Plant Probabilistic Risk Assessment. In *Proceedings of a Symposium on The Assessment and Control of Major Hazards,* I.Chem.E. Symposium Series No 93, Rugby, 1985, pp. 353 - 366.

16. Grether, W.F., In *Psychological Research on Equipment Design,* ed., P.M. Fitts, Army Air Forces Aviation Psychology Program research Report 19, Aero-Medical Laboratory, Wright Field, Ohio, 1947.

17. Warm, J.S., ed., *Sustained Attention in Human Performance,* John Wiley and Sons, Chichester, 1984.

18. Noro, K., Constructing a search model to predict per cent correct counts for each sample. In *Proceedings of the Human Factors Society - 24th Annual Meeting,* Santa Monica, California, 1980, pp. 660 - 664.

19. Wentworth, R.N. and Buck, J.R., Presentation Effects and Eye-Motion, Behaviors in Dynamic Visual Inspection, *Human Factors,* 1982, 24 (6), 643 - 658.

20. Halstead-Nussloch, R. and Granda, R.E., Message-based screen interfaces: The effects of presentation rates and task complexity on operator performance. In *Proceedings of the Human Factors Society - 28th Annual Meeting,* Santa Monica, California, 1984, pp. 740 - 744.

21. Conrad, R., Speed Stress. In *Human Factors in Equipment Design,* eds., W.F. Floyd and A.T. Welford, Lewis, London, 1954.

22. Elkin, E.H., *Effect of scale shape, exposure time and display complexity on scale reading efficiency,* USAF Wright Air Defense Center, Technical Report 58-472, February 1959.

23. Kidd, J.S., A comparison of one-, two- and three-man work units under various conditions of workload. *Journal of Applied Psychology,* 1961, 45 (3), 195 - 200.

24. Miller, G.A., Heise, G.A. and Lichten, W., The Intelligibility of Speech as a Function of the Context of the Test Materials. *Journal of Experimental Psychology,* 1951, 41, 329 - 335.

25. Sumby, W.H. and Pollack, I., Visual Contribution to speech intelligibility in noise. *Journal of the Acoustical Society of America,* 1954, 26, 212 - 215.

26. Jenkins, W.O., The Tactual Discrimination of Shapes for Coding Aircraft-Type Controls. In *Psychological Research on Equipment Design,* ed., P.M. Fitts, Army Air Forces Aviation Psychology Program research Report 19, Aero-Medical Laboratory, Wright Field, Ohio, 1947.

27. Loveless, N.E., Signal Detection with simultaneous visual and auditory presentation. *Flying Personnel Research Committee, Report No 1027,* Air Ministry, London, December 1957.

28. Moffitt, K., Rogers, S.P. and Cicinelli, J., Naming Colours on a CRT Display in Simulated Daylight. In *Proceedings of the Human Factors Society - 30th Annual Meeting,* Santa Monica, California, 1986, pp. 159 - 163.

29. Bradley, J.V., Desirable Control-Display Relationships for Moving-Scale Instruments. *WADC Technical Report 54-423,* Wright-Patterson AFB, Wright Air Development Center, Ohio, September 1954.

30. Collins, B.L., and Lerner, N.D., Assessment of Fire-Safety Symbols. *Human Factors,* 1982, 24, 75 - 84.

31. Conrad, R., Short Term Memory Factor in the Design of Data-Entry Keyboard an interface between short-term memory and S-R compatibility. *Journal of Applied Psychology,* 1966, 50 (5), 353 - 358.

32. Garnham, F., An Alphabetic Alternative to Flashing Light Codes for Telecommunications Equipment. *Post Office Telecommunications Research Department Report 672,* Martlesham Heath, 1979.

33. Vince, M.A., Direction of movement of machine controls. *Flying Personnel Research Committee, Report No 637,* Air Ministry, London, August 1945.

34. Smith, G. and Weir, R., Laboratory Visibility Studies of Directional Symbols used in Traffic Control Signals. *Ergonomics,* 1978, 21, 247 - 252.

35. Webb, R.D.G., Martin, J. and Fisher, W., Response Stereotypes for Rotary Controls involving the application of force. *Perceptual and Motor Skills,* 1982, 55, 275 - 279.

36. Zwaga, H.J. and Boersema, T., Evaluation of a set of graphic symbols. Applied Ergonomics, 1983, 14, 43 – 54.

37. Fitts, P.M. and Seeger, C.M., S–R Compatibility: Spatial Characteristics of Stimulus and response codes. Journal of Experimental Psychology, 1953, 46, 199 – 210.

38. Vermeulen, J., Effects of Functionally or Topographically Presented Process Schemes on Operator Performance. Human Factors, 1987, 29 (4), 383 – 394.

39. Chapanis, A. and Mankin, D.A., Tests of Ten Control–Display Linkages. Human Factors, 1967, 9 (2), 119 – 126.

40. Clatworthy, S.D., Teach Pendants for Industrial Robots: An Investigation into the comfort and safety aspects of their design. Final Year Project Report, Department of Human Sciences, University of Technology, Loughborough, 1983.

41. Warrick, M.J., In Psychological Research on Equipment Design, ed., P.M. Fitts, Army Air Forces Aviation Psychology Program research Report 19, Aero–Medical Laboratory, Wright Field, Ohio, 1947.

42. Carter, L.F. and Murray, N.L., In Psychological Research on Equipment Design, ed., P.M. Fitts, Army Air Forces Aviation Psychology Program research Report 19, Aero–Medical Laboratory, Wright Field, Ohio, 1947.

43. Humphries, M., Performance as a Function of Control–Display Relations, Positions of the Operator and Locations of the Control. Journal of Applied Psychology, 1958, 42, 5, 311 – 316.

44. Ross, S., Shepp, B.E. and Andrews, T.G., Response Preferences in Display–Control Relationships. The Journal of Applied Psychology, 1955, 39, 6, 425 – 428.

45. Spragg, S.D.S., Finck, A. and Smith, S., Performance on a Two-dimensional following Tracking Task with Miniature Stick Control as a function of control–display movement relationships. The Journal of Psychology, 1959, 48, 247 – 254.

46. Colquhoun, W.P., Simultaneous Monitoring of a number of auditory sonar outputs. In Vigilance, Theory, Operational Performance and Physiological Correlates, ed., R.R. Mackie, Plenum Press, 1977, pp. 163 – 188.

47. Conrad, R., Some effects on performance of changes in perceptual load. Journal of Experimental Psychology, 1955, 49, 313 – 322.

48. Kennedy, R.S. and Coulter, X.B., Research Note: The Interactions Among Stress, Vigilance and Task Complexity. Human Factors, 1975, 17, 106 – 109.

49. Mackworth, N.H. and Mackworth, J.F., Visual Search for successive decisions. British Journal of Psychology, 1958, 49, 210 - 221.

50. Poulton, E.C., Two Channel Listening. Journal of Experimental Psychology, 1953, 46, 91 - 96.

51. Poulton, E.C., Measuring the order of difficulty of visual motor tasks. Ergonomics, 1958, 1, 234 - 239.

52. Tickner, A.H. and Poulton, E.C., Monitoring up to 16 synthetic television pictures showing a great deal of movement. Ergonomics, 1973, 16, 381 - 401.

53. Kanarick, A.F. and Petersen, R.C., Effects of value on the monitoring of multi-channel displays. Human Factors, 1969, 11, 313 - 320.

54. Gardner, J.F., The Effect of Motion Relationship and Rate of Pointer Movement on Tracking Performance. WADC technical Report 57-533, ASTIA Document No 131002, Wright Air Development Center, Ohio, September 1959.

55. Vince, M.A., Learning and Retention of an Unexpected Control-Display Relationship under Stress Conditions. Medical Research Council Applied Psychology Unit 125/50, August 1950.

56. Schurick, J., The Effects of Shape Coding and Mirror Imaging of Control Knobs. In Proceedings of the Human Factors Society - 25th Annual Meeting, 1981, pp. 440 - 444.

57. Downing, J.V. and Sanders, M.A., The Effect of Panel Arrangement and Locus of Attention on Performance. Human Factors, 1987, 29 (5), 551 - 562.

58. Fleming, R.A., The Processing of Conflicting Information in a Simulated Tactical Decision-Making Task. Human Factors, 1974, 12 (4), 375 - 386.

59. Brems, D.J., Risk Estimation for Common Consumer Products. In Proceedings of the Human Factors Society - 30th Annual Meeting, Santa Monica, California, 1986, pp. 556 - 560.

60. Karnes, E.W., Leonard, S.D. and Rachwal, G., Effects of Benign Experiences on the Perception of Risk. In Proceedings of the Human Factors Society - 30th Annual Meeting, Santa Monica, California, 1986, pp. 121 - 125.

61. Wogalter, M.S., Godfrey, S.S., Fontenelle, G.A., Desailniers, D.R., Rothstein, P.R. and Laughey, K.R., Effectiveness of Warnings. Human Factors, 1987, 29 (5), 599 - 612.

62. Thompson, S.J., Fraser, E.J. and Howarth, C.I., Driver Behaviour in the presence of child and adult pedestrians. Ergonomics, 1987, 28 (10), 1469 - 1474.

63. Cohen, J., Dearnaley, E.J. and Hansel, C.E.M., Measures of Subjective Probability: Estimates of Success in Performance in Relation to Size of Task. British Journal of Psychology, 1957, 46, 271 - 275.

64. Brown, I.D., Tickner, A.H. and Simmonds, D.C., Interference between concurrent tasks of Driving and Telephoning. Journal of Applied Psychology, 1969, 53, 419 - 424.

65. Long, D.A., Human Factors Considerations for maintenance and repair of off-road haulage trucks. In Trends in Ergonomics/Human Factors III, Proceedings of the Annual International Industrial Ergonomics and Safety Conference, Louisville, Kentucky, USA, 12 - 14 June 1986, Part A, ed., W. Karwowski, North Holland, 1986, pp. 465 - 473.

66. Karnes, E.W. and Leonard, S.D., Consumer Product Warnings: Reception and Understanding of Warning Information by Final Users. In Trends in Ergonomics/Human Factors III, Proceedings of the Annual International Industrial Ergonomics and Safety Conference, Louisville, Kentucky, USA, 12 - 14 June 1986, Part B, ed., W. Karwowski, North Holland, 1986, pp. 995 - 1003.

67. Strawbridge, J.A., The Influence of Position, Highlighting and Imbedding on Warning Effectiveness. In Procedings of the Human Factors Society - 30th Annual Meeting, Santa Monica, California, 1986, pp. 716 - 720.

68. Evans, A.G., Operator Feedback and Touch Controls. Unpublished MSc. dissertation, Ergonomics Department, Cranfield Institute of Technology, and Ranx Xerox Engineering Group, Welwyn Garden City, K78-01261, September 1978.

69. Pollard, D. and Cooper, M.B., The effect of feedback on keying performance. Applied Ergonomics, 1979, 10, 194 - 200.

70. Roe, C.J., Muto, W.H. and Blake, T., Feedback and key discrimination on membrane keypads. In Proceedings of the Human Factors Society - 28th Annual Meeting, 1984, pp. 277 - 281.

71. Castaneda, A. and Lipsitt, L.P., Relation of Stress and Differential Position Habits to Performance in Motor Learning. Journal of Experimental Psychology, 1959, 57, 1, 25 - 29.

72. Monk, A., Mode errors: a user-centred analysis and some preventative measures using keying-contingent sound. International Journal of Man-Machine Studies, 1986, 24, 313 - 327.

73. Armbruster, A. and Fradrich, J., Examination of the influence of the force-distance characteristics and the keyboard slope on the keying error frequency. In Proceedings of the Fourth International Symposium on Human Factors in Telephony, Berlin, 1970, pp. 200 - 212.

74. Long, J., Effects of Delayed Irregular Feedback on Unskilled and Skilled Keying Performance. Ergonomics, 1976, 19/2, 183 - 202.

75. Kao, H.S.R. and Smith,K.U., Unimanual and Bimanual Control in a Compensatory tracking Task. *Ergonomics,* 1978, 21/19, 661 – 670.

76. Deininger, R.L., Human Factors Engineering Studies of the Design and Use of Pushbutton telephone sets. *The Bell System Technical Journal,* 1960, 996 – 1012.

77. Long, J., Visual Feedback and Skilled Keying: Differential Effects of Masking the Printed Copy and the Keyboard. *Ergonomics,* 1976, **19/1,** 93 – 110.

78. Spiker, V.A., Harper, W.R. and Hayes, J.F., The Effect of Job Experience on the Maintenance Proficiency of Army Automotive Mechanics. *Human Factors,* 1985, **27** (3), 301 – 311.

79. Holding, D.H., Repeated Errors in Motor Learning. *Ergonomics,* 1970, **13/6,** 727 – 734.

80. Evans, W.A. and Courtney, A.J., An Analysis of Accident Data for Franchised Public Buses in Hong Kong. *Accident Analysis and Prevention,* 1985, 17/5, 355 – 366.

81. Anderson, J.R. and Jeffries, R., Novice LISP Errors: Undetected Losses of Information from Working Memory. *Human-Computer Interaction,* 1985, **1,** 107 – 131.

82. Widowski, D. and Eyferth, K., Comprehending and Recalling Computer Programs of Different Structural and Semantic Complexity by Experts and Novices. In *Human Decision making and manual control,* ed., H.-P. Willumeit, Elsevier Science Publishers B V (North Holland), 1986, pp. 267 – 275.

83. Behan, R.A., An experimental investigation of the interaction between problem load and level of training. *Human Factors,* 1961, **3,** 53 – 59.

84. Kammann, R., The Comprehensibility of Printed Instructions and the Flowchart Alternative. *Human Factors,* 1975, 17, 183 – 191.

85. Sakala, M.K., *Effects of Format Structure, Sex and Task Familiarity on the Comprehension of Procedural Instructions,* Department of Industrial Engineering, North Carolina State University, Raleigh, NC, Contract No N68 335-75-112, Final report, Naval Air Systems Command, Washington DC, 1976.

86. Stern, K.R., An Evaluation of Written, Graphics, and Voice Messages in Proceduralised Instructions. In *Proceedings of the Human Factors Society – 28th Annual Meeting,* 1984, pp. 314 – 318.

87. Wright, P., Writing to be understood: Why use sentences? *Applied Ergonomics,* 1971, 2, 207 – 209.

88. Klemmer, E.T., Communication and Human Performance. *Human Factors,* 1962, **4,** 75 – 79.

89. Blaiwes, A.W., Formats for presenting procedural instructions. *Journal of Applied Psychology,* 1974, 59, 683 - 686.

90. Booher, H.R., Relative Comprehensibility of Pictorial Information and Printed Words in Proceduralised Instructions. *Human Factors,* 1975, 17, 266 - 277.

91. Weitz, J., The coding of Airplane Control Knobs. In *Psychological Research on Equipment Design,* ed., P.M. Fitts, Army Air Forces Aviation Psychology Program research Report 19, Aero-Medical Laboratory, Wright Field, Ohio, 1947.

92. Colquhoun, W.P., Evaluation of Auditory, Visual and Dual-Mode Displays for Prolonged Sonar Monitoring in Repeated Sessions. *Human Factors,* 1975, 17, 425 - 437.

93. Klemmer, E.T., Time Sharing Frequency-Coded auditory and visual channels. *Journal of Experimental Psychology,* 1958, 55, 229 - 235.

94. Bradley, J.V., Optimum Knob Crowding. *Human Factors,* 1969, 11 (3), 227 - 238.

95. Seibert, W.F., Kasten, D.F. and Potter, J.R., A study of factors Influencing the Legibility of Televised Characters. *Journal of Society of Motion Pictures and Television,* 1959, 68, 467 - 472.

96. Bhatnager, V., Drury, C.G. and Schiro, S.G., Posture, Postural Discomfort and Performance. *Human Factors,* 1985, 27 (2), 189 - 199.

97. Bartlett, F.C. and Mackworth, N.H., *Planned Seeing,* H M Stationery Office, London, 1950.

98. Giddings, B.J., Alpha-Numerics for Raster Displays. *Ergonomics,* 1972, 15/1, 65 - 72.

99. Mackworth, N.H., Effects of heat on wireless telegraphy operators hearing and recording morse messages. *British Journal of Industrial Medicine,* 1946, 3, 143 - 158.

100. Alluisi, E.A., Beisel, W.R., Bartellonis, P.J. and Coates, G.D., Behavioral Effects of Tularemia and sandfly fever in man. *Journal of Infectious Diseases,* 1973, 28, 710 - 717.

101. Griffiths, J.D. and Boyce, P.R., Performance and Thermal Comfort. *Ergonomics,* 1971, 14, 457 - 468.

102. Wilkinson, R.T., Fox, R.H., Goldsmith, R., Hampton, I.F.G. and Lewis, H.E., Psychological and physiological responses to raised body temperature. *Journal of Applied Physiology,* 1964, 19, 287 - 291.

103. Broadbent, D.E., Some Effects of Noise on Visual Performance. *Quarterly Journal of Experimental Psychology,* 1954, 6/1, 1 - 5.

104. Garcia, K.D. and Wierwille, W.W., Effect of Glare on Performance of a VDT Reading-Comprehension Task. *Human Factors*, 1985, **27** (2), 163 - 173.

105. Weston, H.C., The Relation between illumination and visual efficiency - the effect of brightness contrast. *Industrial Health Research Board Report No 87*, HMSO, Medical Research Council, London, 1945.

106. Benor, D. and Shvartz, E., Effect of body cooling on vigilance in hot environments. *Aerospace Medicine*, 1971, **42**, 727 - 730.

107. Poulton, E.C., *Environment and Human Efficiency*, Charles C. Thomas, Springfield, Illinois, 1970.

108. Bell, P.A., Effects of Noise and Heat Stress on Primary and Subsidiary Task Performance. *Human Factors*, 1978, **20**, 749 - 752.

109. Boyce, P.R., *Human Factors in Lighting*, Applied Science Publishers, London, 1981.

110. Baker, R.A., Ware, J.R. and Sipowicz, R.R., Vigilance: a comparison of auditory, visual and combined audio-visual tasks. *Canadian Journal of Psychology*, 1962, **16**, 192 - 198.

111. Beshir, M.Y., Time-on-Task period for unimpaired tracking performance. *Ergonomics*, 1986, **29**, 423 - 431.

112. Jerison, H.J. and Pickett, R.M., Vigilance: the importance of the elicited observing rate. *Science*, 1964, **143**, 970 - 971.

113. Lisper, H.O., Laurell, H. and Van Loon, J., Relation between time to falling asleep behind the wheel on a closed track and changes in subsidiary reaction time during prolonged driving on a motorway. *Ergonomics*, 1986, **29**, 445 - 453.

114. Mackie, R.R., ed., *Vigilance: Theory, Operational performance and Physiological Correlates*, Plenum Press, New York, 1977.

115. Mackworth, N.H., Researches on the measurement of Human Performance. In *Selected Papers on Human Factors in the Design and Use of Control Systems*, ed., H.W. Sinaiko, Dover Publications, New York, 1961, pp. 174 - 331.

116. Mackworth, J.F., The d' measure of signal detectability in vigilance-like situations. *Canadian Journal of Psychology*, 1963, **17**, 302 - 325.

117. Johnson, E.M. and Payne, M.C., Vigilance: Effects of frequency of knowledge of results. *Journal of Applied Psychology*, 1966, **50**, 33 - 34.

118. Jerison, H.J., On the decrement function in human vigilance. In *Vigilance A Symposium*, eds., D.N. Buckner and J.J. McGrath, McGraw-Hill, 1963, pp. 199 - 212.

119. Harris, W. and Mackie, R.R., A study of relationships among fatigue, hours of service, and safety of operations of truck and bus drivers. Final report. Technical Report 1727-2, Human Factors Research Inc., Goleta, California, 1972.

120. Ball, C., Funk, T., Noonan, D., Velasquez, J. and Konz, S., Degradation of Performance due to Sleep Deprivation: A Field Test. In Proceedings of the Human Factors Society – 28th Annual Meeting, Santa Monica, California, 1984, pp. 570 – 574.

121. Webb, W.B. ed., Biological Rhythms, Sleep and Performance, John Wiley and Sons, Chichester, 1982.

122. Haslam, D.R., Sleep Deprivation and Naps. Behavior Research Methods, Instruments and Computers, 1985, 17 (1), 46 – 54.

123. Johnson, L.C., Sleep deprivation and performance. In Biological Rhythms, Sleep and Performance, ed., W.B. Webb, John Wiley and Sons, Chichester, 1982.

124. Tilley, A.J., Wilkinson, R.T., Warren, P.S.G. and Drud, M., The Sleep and Performance of Shift Workers. Human Factors, 1982, 24, 629 – 641.

125. Giambra, L.M. and Quilter, R.E., A Two-Term Exponential Functional Description of the Time Course of Sustained Attention. Human Factors, 1987, 29 (6), 635 – 643.

126. Colquhoun, W.P., ed., Aspects of Human Efficiency, Proceedings of a Conference held at Strasbourg in July 1970 under the aegis of the NATO Scientific Affairs Division.

127. Colquhoun, W.P. and Rutenfrantz, J., eds., Studies of Shiftwork, Taylor and Francis, London, 1980.

128. Levine, J.M., Samet, M.G. and Brahlek, R.E., Information seeking with input pacing and multiple decision opportunities. Human Factors, 1974, 16, 384 – 394.

129. Coury, B.G. and Drury, C.G., The Effects of pacing on complex decision-making inspection performance. Ergonomics, 1986, 29, (4), 489 – 508.

130. Schlegal, B. and Beneke, M., A Study of Self-Paced and Machine-Paced Inspection. In Proceedings of the Human Factors Society – 30th Annual Meeting, Santa Monica, California, 1986, pp. 471 – 475.

131. Klemmer, E.T. and Stocker, L.P., Effects of Grouping of Printed Digits on Forced-Paced Manual Entry Performance. Journal of Applied Psychology, 1974, 59 (6), 675 – 678.

132. Peddada, T. and Bennett, C. A., Inspection Contrasting Self-Pacing and Machine-Pacing. In *Proceedings of the Human Factors Society - 28th Annual Meeting*, Santa Monica, California, 1984, pp. 675 - 677.

133. Waag, W.L. and Halcomb, C.G., Team size and decision rule in the performance of simulated monitoring teams. *Human Factors*, 1972, 14, 309 - 314.

134. Weiner, E.L., The performance of multi-man monitoring teams. *Human Factors*, 1964, 6/2, 179 - 184.

241

MODELLING THE EVACUATION OF THE PUBLIC IN THE EVENT OF TOXIC RELEASES: A DECISION SUPPORT TOOL AND AID FOR EMERGENCY PLANNING

PAUL I. HARRISON and LINDA J. BELLAMY
Technica Ltd., Lynton House, 7/12 Tavistock Square, London WC1H 9LT

INTRODUCTION

This paper presents some of the results of a study undertaken for H.M. Nuclear Installations Inspectorate by Technica to assess the factors which would affect the times taken to evacuate the civil population in the vicinity of nuclear power stations. The aims of the study were to:

(a) Renew the evacuation data base and develop a model of evacuation

(b) Estimate total evacuation times for specified populations

and

(c) Consider any implications for Emergency Planning.

it will be shown that the developed model forms the basis of a useful decision support tool for Emergency Planning.

MODEL DEVELOPMENT

As with previous work conducted by Technica, the evacuation models assessed had deficiencies making them difficult to apply in this context. As a result a model of the evacuation process was developed which could be used predictively.

Essentially the final model used was an amalgamation of two existing models of the evacuation process. The first was one developed by Technica for a previous study on evacuation from toxics (VROM study, Technica 1986). This model, currently being used in fire research, describes the behavioural stages that occur in response to a threat, and which result in a behavioural response(s) such as evacuation or non-evacuation.

Very simply described, once a decision has been made to warn a population and the warning process begins, then on an individual level, people will make their behavioural response on the basis of the level of threat they perceive. If a sufficiently high perception of threat exists then evacuation may be the resulting behavioural response choice. People may alternatively decide to seek additional information and adjust their level of perceived threat accordingly - either to then evacuate or not.

The stages in this model were then represented within discrete time components each of which take a certain time to execute. These time components (as used by Urbanik, 1980) are:

Decision Time: This starts from the moment a threat is identified and continues to the point at which it is considered serious enough to warrant issuing a public warning.

Notification Time (NT): This starts from the point at which it is recognised that a warning must be given to the public and ends when 100% (theoretically) of the target population has received the intended warning. (Of course, not all of a population may receive a warning.)

Preparation Time (PT): This commences from the point at which notified individuals decide to evacuate, and prepare to effect such a response. It includes such behaviour as obtaining information on reception areas, determining the routes to take (where options are available), waiting for official transportation and collecting family members who are not present. Some people could of course be evacuating while others are still being notified or preparing to leave.

Evacuation Time (ET): This starts once the process of evacuation (actually leaving) occurs. It includes the time taken to exit the evacuation zone.

This Model is shown in Fig.1.

The time distributions of the NT, PT and ET components in the model will overlap one another. This is because:

- The entire population cannot be notified simultaneously.

- There will be variations in decision times for making a response (if any), with individuals waiting for or seeking additional information.

- There will be variations in preparation time.

Thus, evacuation will commence at different times for different members of the population.

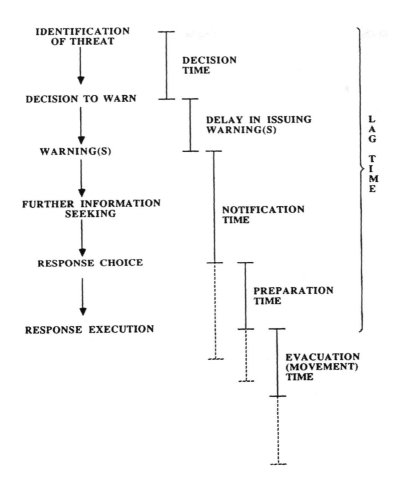

FIGURE 1. EVACUATION TIME MODEL COMPONENTS

Note that there will be a certain amount of time required after deciding to issue a warning, before the first people begin to evacuate. This time taken to get the evacuation process into the actual movement phase is usefully described as the "lag" time. After this lag time, people will begin to evacuate at a certain rate i.e. people per hour.

244

DATA REQUIREMENTS

According to this final model, it was then necessary to analyse Technica's evacuation data base in order to, firstly, determine the variables that affect each time component, and secondly, to record the time ranges that have occurred in the execution of each component. Data on evacuations from toxics and radiological hazards were used.

The variables that were found to affect each time component are shown in table 1.

A discussion of some of these variables is given in a following section which looks at how such information may be used in Emergency Planning.

TABLE 1
Time components and their variables

DECISION TIME	OFFICIAL NOTIFICATION & PUBLIC'S DECISION TIME	PREPARATION TIME	EVACUATION TIME
Threat characteristics/ accident scenario Measurement resources	Warning media Warning content Sources of warnings Additional information acquisition Perception of threat (e.g. attitudes to nuclear power) Location of sectors of population Age characteristics of population	Children in family Location of members of household (together or not) Chosen destinations Instructions from authorities Intended mode of evacuation (e.g. official buses, private cars etc.) Work obligations (e.g. farm vs. home)	Availability of escape routes Resources for escape Resident and transient population numbers

NB. All variables listed are taken from evidence in the literature. Other variables may obviously have an effect (e.g. time of day on the notification time), but no direct evidence has been found for these.

TABLE 2
Summary of time ranges of components of the evacuation model

COMPONENT	RANGE
Decision Time	1-52 hours
Notification Time	
i) Delay time	1-48 hours
ii) Warning time	1-48 hours
Preparation Time	1-24 hours
Evacuation (movement) time (shown as a rate)	113-11,100 persons per hour

Evacuation rate was derived by taking the total numbers evacuated for an evacuation incident, and dividing by the total time from first people moving after an official warning, to the time at which all evacuating persons had evacuated. For example if 2000 people had been evacuated in 2 hours, then a rate of 1000 people per hour is recorded as the evacuation rate achieved for that particular incident.

However, where the data did not allow identification of the time at which first people started moving, then the time from first official warnings was used to derive rate. Also, where evacuations were carried out in stages, or where some sectors of the population may have been kept waiting for official transport to a safe haven, there would inevitably be some adding in of dead time.

Once the evacuation itself has started, it appears that the rate is dependent on the numbers evacuating.

Using the data in Technica's toxic incidents evacuation data base, the available U.K. data points, and also the data for nuclear incidents the points were plotted of rate against number evacuated on a log-log scale. It should be noted that the axes were drawn in such a way as to imply that rate of evacuation is dependent on numbers evacuating. It is possible, however, that the rate reflects resources availability and that this is the limiting factor for the numbers evacuating. However, as the number evacuating is the known variable, we have taken the rate as the dependent variable. We also have no evidence that we should accept the alternative.

A line of best fit was plotted using all of these data together (see Figure 2). This was done by using the method of least squares. In order to highlight the spread in the data, $\pm$ 1 standard deviation from the mean line is shown. 68% of the data fall between these lines. It should be noted that 75% of the data used in the rates graph involved evacuating populations of less than 9000 people, and that little data exist for very large population evacuations.

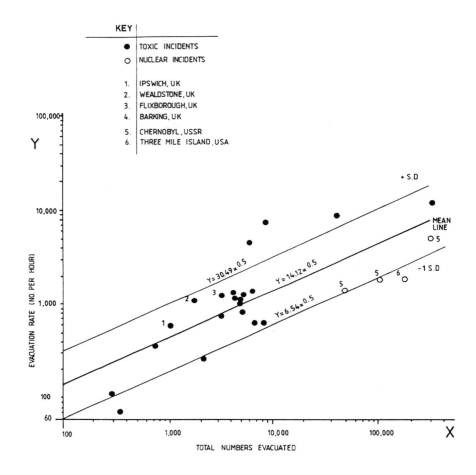

FIGURE 2. LINE OF BEST FIT THROUGH EVACUATION DATA POINTS

In our provisional evacuation time estimates, rates have been derived using the mean line.

Evacuation Rate is then given by:

$y = 14.12 \ (x)^{0.5}$, where;

x = the numbers to be evacuated
y = the evacuation rate (numbers per hour).

At the end of this preliminary analysis it was possible to state which factors affect the evacuation rate, and also factors that broadly influence the way in which the stages overlap.

This knowledge was then used in the formulation of a method to estimate total evacuation times.

METHOD USED TO ESTIMATE TOTAL EVACUATION TIMES

The task was to estimate total evacuation time for specific populations e.g. within a 5-mile radius quadrant about a power station.

Using the understanding gained of the factors that affect the evacuation process, it was possible to gather site specific data on these factors. In order to illustrate this method, a hypothetical example area will be used and a description of the procedure for calculating evacuation time given.

Procedure for Time Calculations

The procedure by which the time estimates are made is documented in the steps outlined below:

1. Divide area around site into sectors and radii of interest.

   Note that it is usual to divide an area into 30^0 sectors for the purposes of emergency planning. Although one would not expect a plume spread of 90^0, we consider that quadrants are more appropriate for modelling nuclear accidents due to the evacuation shadow phenomenon. This is where far more people tend to evacuate than are warned.

2. Determine permanent population numbers for each sector and radius.

3. Determine transient population numbers from information provided and/or by examining maps for the locations of tourist attractions and factoring up the permanent population numbers.

248

4. Add permanent and transient population numbers for each sector and radius. This gives the values of x to be inserted into the rates equation.

5. Makes an estimate of any delay that might be expected after the decision to warn the population and before notification actually begins.

6. Evaluate notification time for each sector and radius.

Note that the availability of actual data is limited here. For the U.K. we would assume that beyond any zone where an emergency plant exists the main method of notification would be by the media. Otherwise we assume the use of tannoys, sirens and door-to-door knocking. A rough guide is shown in Table 3 for 90° sectors.

TABLE 3
Scale for estimating notification times for 90° sectors

RADIUS	TIME TO COMPLETE NOTIFICATION (hrs)
^{1}/2 mile	2 hours
2 miles	3 hours
5 miles	4 hours
10 miles	6 hours
15 miles	7 hours
20 miles	8 hours

7. Evaluate preparation time for each sector and radius.

Again, the availability of real data is limited and highly dependent upon the length of time of effect, imminence of threat and any necessity for shutting down farms, factories etc. Some gross estimates are shown in Table 4.

249

TABLE 4
Scale for preparation times

TIMES (hrs)	COMMENT
1	No farms and institutions
2	*Family effect, no farms or institutions
3	Farms and/or institutions

* See later discussion on the effects of the variables

8. Calculate evacuation rate using the rates equation and the population calculated in Step 4. It may be necessary to adjust the population estimates to account for non-evacuees. This is more important when small scale evacuations are being evaluated.

 The decision as to whether to use the mean or standard deviation lines will depend on the scenario. For U.K. toxic accidents it is recommended that the +1 S.D. line is used. Lack of U.K. data for nuclear accidents suggests use of the mean line unless otherwise indicated.

9. Divide the population number by the rate calculated in Step 8 to obtain evacuation (movement) time estimates.

10. Evaluate the extent of overlap of the stages to obtain the lag time. Obviously, shorter lag times will result from rapid implementation of an effective emergency plan, with a "trained" population, who should evacuate with little or no delay. This may contrast to a poor plan, delays in the issue of warnings, or in waiting for special transport. A range of 30-150 mins is considered appropriate, with 30 minutes being used for small scale evacuations in the U.K. (few hundred people) where emergency plans exist and the population is in imminent danger.

11. By adding notification time and preparation time the theoretical maximum time to get the entire population on the move is estimated.

12. Lag time, plus evacuation (movement) time calculated from the rates equation, gives the total time to evacuate the entire sector radius.

250

EXAMPLE CALCULATION

A hypothetical example is used to demonstrate the model. A map of the area showing the radii and sectors around an imaginary nuclear power plant is given in Figure 3. The coastline has many sandy beaches and offers considerable attractions to tourists in this and surrounding areas. Beyond 5 miles of the plant there are farming areas.

The shape of the coastline is such that the plant is sited on a peninsula fed by a single A road. This feeds into a motorway.

In making the time estimates the decision making phase was not considered. Delay after the decision to evacuate was estimated as negligible for this example.

An estimate of lag time was 2 hours for the 2 mile radius and beyond, and 1 hour for areas within $1/2$ mile because here an emergency plan exists. These lags would have been shorter if a toxic or flammable incident were being considered.

The results are shown in Table 5. It can be seen that evacuation is predicted to take between 5-44 hours depending on quadrant and radius for a mean rate of evacuation.

Rates can then be compared with escape route capacity. If, for example, road capacity does not allow the calculated rate of evacuation, then movement time should be increased accordingly. The peninsula should clearly be examined as this has only a single escape road. By estimating the population figures just for the promentary, within the 5 mile radius, a rate of evacuation can be calculated for this area. The population data for the SW and NW sectors were added and the value of 78,463 persons put into the mean rates equation. This gives an evacuation rate of 3955 persons per hour.

If we assume 2 persons per car, then the rate of evacuation is 1977 cars per hour. Say the flow rate capacity of the hypothetical A85 is 2500 cars per hour; by dividing the road capacity by the evacuation rate, then:

$$\frac{2500 \text{ cars/hr}}{1977 \text{ cars/hr}} = 1.3$$

In other words, the escape system is only just able to cope with the estimated evacuation rate.

Finally, if one uses the +1 S.D. formula (where appropriate), an evacuation rate of 8541 persons/hr or 4270 cars per/hr is obtained. Then:

$$\frac{\text{Road flow rate}}{\text{Evacuation rate}} = 0.6$$

Figure 3 : Map of Area Around a Hypothetical NPP Site

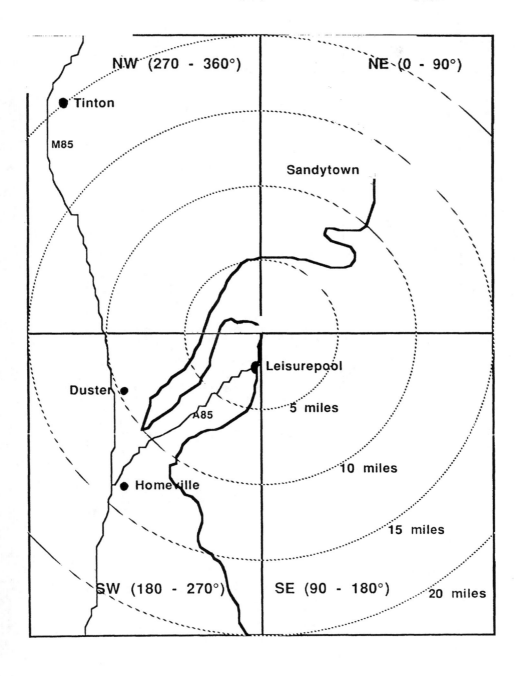

QUADRANT (degrees from N)	DISTANCE (RADIUS)	PERMANENT POPULATION ESTIMATES	TRANSIENT POPULATION ESTIMATES	TOTAL POPULATION ESTIMATES	MAX. TIME TO GET POPULATION ON THE MOVE NOTIFICATION TIME (Hrs) DELAY AFTER DECISION TO EVACUATE	ISSUING WARNING TO 100% OF POPULATION	MAX. PREPARATION TIME (Hrs)	LAG TIME DERIVED FROM OVERLAP (Hrs)	MOVEMENT TIME MEAN (Hrs)	MOVEMENT TIME RANGE (Hrs)	AVERAGE TIME TO CLEAR ZONE TOTAL EVACUATION TIME (Hrs) (LAG TIME + MEAN MOVEMENT TIME)
180-270 (SW) Duster, Homeville, Leisurepool	$^1/_2$ mile		3075	3075	0	2	1	1	4	1- 5	5
	2 miles	5734	4625	10359	0	3	2	2	7	3-12	9
	5 miles	54767	+ 25%	71200	0	4	3	2	19	8-39	21
	10 miles	98900	+ 10%	108790	0	6	3	2	23	11-50	25
	15 miles	118821	+ 10%	130703	0	7	3	2	26	12-55	28
	20 miles	127505	+ 10%	140256	0	8	3	2	26	12-59	28
270-360 (NW) Sparsley populated, but with edge of Tinton	$^1/_2$ mile		3075	3075	0	2	1	1	4	1- 7	5
	2 miles	530	4625	5155	0	3	2	2	5	2- 8	7
	5 miles	2421	+200%	7263	0	4	3	2	6	2-10	8
	10 miles	13133	+ 45%	19043	0	6	3	2	10	4-20	12
	15 miles	29063	+ 25%	36329	0	7	3	2	13	6-27	15
	20 miles	66417	+ 15%	76380	0	8	3	2	20	9-41	22
0-90 (NE) Sandytown area	$^1/_2$ mile										
	2 miles				NOTHING BUT SEA						
	5 miles										
	10 miles	22373	+ 20%	26848	0	4	3	2	12	5-24	14
	15 miles	153005	+100%	306606	0	7	3	2	39	14-65	41
	20 miles	232506	+ 50%	348759	0	8	3	2	42	17-77	44
90-180 (SE)					NOTHING BUT SEA						

TABLE 5
Evacuation Time Estimates

Thus, the evacuation rate is no longer within the escape route capacity. This is a potential condition for panic, where people may feel unable to escape from what they perceive as a high threat situation.

IMPLICATIONS FOR EMERGENCY PLANNING

The model should be seen as a tool for use in Emergency Planning. It is useful in two main areas. The first being an aid for decisions as to whether evacuation or sheltering should occur, and the second, the extent to which the effects of the variables in the model are considered in planning. These are now discussed in turn.

Evacuate of Shelter?

If the total evacuation time predicted by the model exceeds the available time before injurious effects on a population - then sheltering should be considered in place of evacuation.

Alternatively, a decision to evacuate may have to be made at as early a point in the decision stage as possible/necessary, in order to allow enough time for the evacuation to occur, before the critical conditions are reached.

Indeed, it may become necessary to adopt a policy of sheltering sections of the population while evacuating the most endangered instead of attempting a total evacuation in one go. For radiological releases however, such a philosophy may be unpracticle due to the evacuation shadow phenomenon. This is where far more people actually evacuate than are required to do so. It is a phenomenon peculiar to radiological threats, and contrasts with toxic threats requiring evacuation, where fewer people tend to evacuate than are warned.

The Effects of the Variables and the Implications for Emergency Planning

The variables that are listed in Table 1 were all found to have an effect on the evacuation process. A study of how they may actually exert their effect has implications for the planning process.

For example, consider warning content as a variable affecting the notification time. To be most effective, a warning should be unambiguous:

       i.e. - convey sense of danger
            - convey imminence of the threat
            - give instructions on the appropriate response to the
              threat

Ambiguous warning content may cause longer notification times. It is important therefore to consider this concept when deciding on the content of a warning.

A variable affecting preparation time, is the location of members of the household. Families act as units in a crisis, and if members are apart then they may wait until they are united before evacuating. It is common practice in an emergency to set up an exclusion zone around the danger area, and in this way prevent people entering the zone. This may have adverse effects unless the evacuating population is told that their spouse/relatives will not be able to re-unite since the exclusion zone prevents this. They should also be told that they should, for example, meet relatives at a designated meeting point outside the exclusion zone. The provision of such information may allow more rapid evacuation by reducing preparation time.

More obvious perhaps is the effect of the capacity of escape routes on Evacuation (movement) time. Bottleneck areas may cause increased movement times, and the extent of such bottlenecks will determine the level of traffic management that is required to ensure rapid evacuation.

INFORMATION PROVISION FOR POPULATIONS IN "DESIGN ACCIDENT" ZONES

It is certainly policy in the U.K. now to provide households within the areas considered to lie in design accident radii, with a relevant information leaflet. This is designed to "train" the population such that in the event of a release they are more familiar with both the actions of the authorities, and the required personal response. This is a very important aspect in emergency planning, and because of this it is suggested that some standards be established for such documents to increase the probability that information is remembered and used "on the day". Appropriate Psychological and Ergonomic research should be used when determining both the content and layout of information.

FUTURE DATA COLLECTION

It would be useful to expand the data base on evacuations and therefore helpful if data could be collected in a format compatible with the evacuation model's requirements. A more extensive and structured collection of data would further our understanding of the evacuation process and aid predictions of total evacuation time.

It is suggested then that during an evacuation incident estimates be made for all the component times within the model, and recordings made of the main factors that affected the stage times, the lag time, and also the relative overlap of stages.

MODEL VALIDATION

This simple model of the evacuation process has been tested out on two previous evacuations that have occurred in the past.

This benchmark study involved a "blind" investigator (i.e. with no knowledge of the times that occurred apart from accident onset) asking questions relevant to the model's requirements, and estimated the total evacuation time.

In both cases, the predicted total evacuation times were accurate within $1/4$ hour of the actual evacuation times.

Although simple, the model would appear to be a useful tool for Emergency Planning.

REFERENCES

[1] Technica (1986) Review of Evacuation Data, September 1986. Report prepared for the Ministry of the Environment (VROM) Netherlands.

[2] Technica (1987a) Evacuation of the Public in the Vicinity of Nuclear Installations. Vol. I. A Review of Factors Affecting Evacuation, July 1987. Report prepared for H.M. Nuclear Installations Inspectorate, U.K.

[3] Technica (1987b) Evacuation of the Public in the Vicinity of Nuclear Installations. Vol. II. Specific Estimates of Time to Evacuate from Two Sites. Report prepared for H.M. Nuclear Installations Inspectorate, U.K.

[4] Urbanik, T. (1980) Analysis of Techniques for Estimating Evacuation Times for Emergency Planning Zones. NUREG/CR/-1745.

Technica Ltd, acknowledge the permission of HM Nuclear Installation Inspectorate to use material from a study undertaken for HM NII. The opinions expressed are those of the Authors.

DESIGNING DECISION SUPPORT SYSTEMS FOR HUMAN ERROR REDUCTION: THE NEED TO ADDRESS INFORMATION DISTORTION

WILLIAM L. CATS-BARIL
School of Business Administration
University of Vermont
Burlington, Vermont 05405 USA

ABSTRACT

This article proposes that the principle behind the design of Decision Support Systems should be the minimization of distortion in the collection, processing, and transmission of information. After suggesting that suboptimal decision-making is a result of both cognitive biases and faulty communication of information, a framework which matches sources of cognitive and organizational communication distortion with strategies to neutralize them is presented. The article concludes by offering a design methodology and an example on how to apply it.

INTRODUCTION

Individuals in organizations are expected to make decisions on the basis of information collected, processed and transmitted through a variety of channels. The reliability of these channels is seldomly known and difficult to assess because of the large number of factors that may impact that reliability. For example, assume that individual "A" has to make a decision partially based on information provided by individual "B". The content of the information that "B" will transmit to "A" depends to a large extent on the perceptions that "B" has of what the needs of "A" are, and his perceptions of the importance, relevance, timeliness, and impact of the message he is transmitting to "A". Furthermore, these perceptions are colored by status differences between "A" and "B", their relative position in the organizational structure, differences in training, cultural background, and cognitive style among other factors.

From an organizational decision making point of view, the transmission of information and its potential distortion is only one of two concerns. The other concern is the existence of a myriad of heuristics, biases and psychological characteristics which decrease human performance in every phase of the decision process: from problem recognition to designing solutions to interpreting feedback. Indeed, a review of the cognitive psychology literature establishes very clearly that decision makers need substantial help in information processing [1]. What this means is that even if individuals were to be provided with distortion-free information, they are likely to process it suboptimally.

Moreover, managerial work is characterized by time pressures, frequent interruptions, divided attention among a wide variety and high number of tasks, and high stakes [2]. These circumstances make the potential impact and incidence of the cognitive and organizational sources of distortion described above so much more likely and acute.

The premise of this article is that Decision Support Systems (DSS) should be designed to provide relevant and undistorted information for decision-making. This premise is implemented by focusing on minimizing the probability of distorting the needed information as (i) it is communicated throughout the organization, and (ii) it is processed by a specific individual.

The implication is that the traditional DSS development cycle -- with its emphasis on prototyping and information requirements determination (IRD) -- should be broadened to consider the sources of distortion that may have an impact on the quality of decision making. The methodology presented here takes the typical IRD phase and expands it to include an analysis of the cognitive biases that may affect the processing of information at the individual level and the potential sources of distortion in the collection and transmission of that information at the organizational level.

A review of the literature on DSS makes clear that while researchers have given thought to the impact of several variables on the design of DSS for some time [3], they have given scant attention to the crucial aspects of human information processing and the reasons why managers exhibit suboptimal decision-making behavior.

Also, the great majority of the literature in this field assumes implicitly that either the data being processed by the decision maker is undistorted or, if it is distorted, then it is "someone-else's-job-to-clean-it-up". That is, in setting the boundary for the design of a DSS and determining the reliability of its information, researchers in this field have seldomly ventured much further than the desk of their target user and the reports that land on it.

This paper is organized in three parts. First, a review of the research findings on information processing distortion due to human information processing and organizational communication is presented. Then, a framework to classify and design DSS is described. The article concludes by providing an example on how to apply the design strategy.

REVIEW

Decision Support Systems (DSS) is a label describing those information systems that aim: 1) to impact on ill-structured decisions where judgement is essential, 2) to support the user instead of replacing him, and 3) to improve the user's effectiveness as a decision maker [4, 5].

In order to fulfill these aims, DSS have to address those factors that lead to suboptimal decision making and make an individual a less effective decision maker. Two major factors contribute to distortions in information processing caused by cognitive biases and the distortions in information brought about by certain characteristics of the organizational communication network. These sources of distortion are discussed below.

Distortion

Cognitive: There are three main reasons to help human beings process information: (1) the limitations of their perceptual system, (2) the pervasiveness of simplistic and biasing strategies affecting the acquisition and processing of information, and (3) the information processing idiosyncracies of individuals due to certain personality characteristics.

The limitations of our perceptual system are well known: information is assimilated at relatively low rates; the processing is sequential and constrained by a low capacity short-term memory; the use of selective perception, habitual schemas, programs and filters hampers the analysis of new tasks. These limitations affect the way we do computations, remember and retrieve information.

The systematic biases that pervade the acquisition and processing of information have also been well documented. For example, Tversky and Kahneman [6] present evidence suggesting that the ease with which specific instances can be recalled from memory affects judgments of frequency of occurrence. They also point out that individuals' judgments are insensitive to sample sizes. Other researchers have found that in assessing frequencies, individuals tend to ignore the non-occurrence of an event, using in their assessment the observed frequency rather than the relative one and ignoring baseline information [7]. Individuals also have a tendency to assign a greater diagnostic value than they should to concrete data based on vivid description of actual incidents.

Human information processing is also affected by the order in which the data is presented. Biases induced by the order of presentation include anchoring, halo, recency, and primacy effects. Other biases include hindsight explanations of unexpected outcomes, misunderstandings of chance fluctuations, illusory correlation, and the use of inappropriate schemas [8].

Personality characteristics like integrative complexity [9], dogmatism [10], risk taking [11] and creativity [12] affect the way people process information. Finally, cognitive style has been proposed as an important variable in the preference for and presentation of information [3].

Stress brought on by environmental complexity (e.g., distractions, time pressures), not knowing how to solve a problem, fatigue, or the magnitude of regret increases the likelihood of these biases [1, 13].

Organizational Communication: So far we have considered sources of distortion which are internal to the individual. We now turn our attention to sources of distortion that are related to the structure of the organization. Building on Huber's classification of organizational information processing [14], distortion is defined as the mis-routing, delaying, and faulty modification and summarization of information.

The quality of decisions is affected not only by the quality of the information processing at the individual level, but also by the validity and reliability of the information he receives through the communication system of the organization. The communication system is the network through which requested and unrequested information flows.

Furthermore, the transmission of information in organizations is far from being smooth. For example, Wiksell [15] claims that 70% of all information communicated in organizations is bound to be distorted, misunderstood, rejected, forgotten, or disliked. MacCrimmon [16] in studying team decision making found that 90% of the policy makers he used as subjects exhibited inefficient and mutually inconsistent communication heuristics.

There is strong evidence that distortion is a function of the status and location of the sender and receiver of a message, the structure of the organization (in particular hierarchies), and the content of the message. For example, Athanassiades [17] in investigating the relationships between distortion of upward communication, needs of subordinates and aspects of organizational climate, found that distortion is negatively correlated to the communicator's level of security and positively correlated to his achievements needs and the degree of authority in the hierarchy. In general, if the sender perceives a high cost, either material or psychological, the probability of distortion is high.

Other findings have linked structural variables to dysfunctional communication. Specialization and differentiation create problems of coordination, uneven perceptions and expectations, and erratic transfer of information. Formalization increases the probability of distortion because more links are added to the communication chain; centralization inhibits large amounts of information at the top from being passed down, and causes information needed at the top to get there late, if at all.

The direction of the message is also correlated with distortion. For example, low-status individuals attempt to communicate (without reciprocation) with individuals higher up in the organization more often than circumstances would require it, and less often than needed with other low-status individuals.

Finally, the degree of distortion is affected by the characteristics of the message itself. The amount of ambiguity, perceived timeliness and relevance, negative (bad news tend to move much more slowly than good news) and dissonant content all have an impact on the likelihood of the message being distorted.

A summary of this discussion is offered in Table 1 .

TABLE 1
Sources of information distortion

SOURCES OF INFORMATION DISTORTION	
COGNITIVE	ORGANIZATIONAL
PHYSIOLOGICAL LIMITATIONS BIASES TRAINING/BACKGROUND PERSONALITY CHARACTERISTICS COGNITIVE STYLE	DIRECTIONALITY POWER AND STATUS TRUST AND EXPECTATIONS STRUCTURE NUMBER OF LINKS

THE FRAMEWORK

As stated before, the DSS literature has emphasized the importance of supporting the individual by increasing his effectiveness as a decision maker. Support, in this article, is defined as those operations of a DSS that reduce distortion in the collection, transmission and processing of information. To determine how much support is being provided by a DSS, we propose to measure how helpful the DSS is in pinpointing, alleviating and/or eliminating sources of distortion in the transmission of information, on the one hand, and in the processing of information on the other. Just as we discussed distortion at two levels, support can also be classified into two types: Cognitive and Organizational.

The goal of Cognitive Support is to address the weaknesses in human information processing by providing a systematic and consistent treatment of available information. The purpose of Organizational Support is to provide timely, relevant and distortion-free information to the individual needing that information to make a decision.

Cognitive Support focuses on alleviating the biases and limitations in individual information processing, while Organizational Support concentrates on designing -- assuring the reliability and timeliness of-- the best communication network that will deliver the needed information. In other words, while Cognitive Support attempts to alleviate biases and limitations of the user, Organizational Support aims at reducing distortion by assuring the reliability and validity of the communication system. The two types of support are compared in Table 2 and discussed in detail below.

TABLE 2
Comparison of Cognitive and Organizational Support

| | TYPE OF SUPPORT | |
	COGNITIVE	ORGANIZATIONAL
FUNCTION	TO HELP INDIVIDUALS PROCESS INFORMATION	TO HELP COLLECT AND TRANSMIT REQUIRED INFORMATION THROUGHOUT THE ORGANIZATION
FOCUS	TO AVOID INCONSISTENCIES IN PROCESSING INFORMATION TO SOLVE A GIVEN TASK	TO PROVIDE RELIABLE INFORMATION THROUGH AN EFFICIENT COMMUNICATION NETWORK
GOAL	TO DEBIAS MENTAL PROCESSES	TO HAVE DISTORTION-FREE COLLECTION AND TRANSMISSION OF INFORMATION

Cognitive Support

Hackathorn [18] called Personal Computing an information processing activity in which the user has direct personal control over all the stages of the activity and one which emphasizes small-scale technology and localized systems. Following Hackathorn, we will refer to Cognitive Support as the support provided by a system whose main purpose is to interact with one user and help him structure problems, check assumptions, develop recommendations, etc.

While Cognitive Support concentrates on improving the decision maker's ability to process information, it does not address the issues of assuring the reliable transmission of information. Cognitive Support can be classified into four categories: (a) indexing, (b) computational support, (c) procedural support, and (d) display/representational support. information).

Indexing support provides the user with data storage and retrieval capabilities. Examples are census data books, stock market data, sales histories, and indexed databases.

TABLE 3

Different types of Cognitive Support and the Cognitive
Limitations They Address

TYPE OF COGNITIVE SUPPORT	COGNITIVE LIMITATION(S) OR SOURCE OF BIAS
INDEXING (supporting the storage and retrieval of data)	. limited memory . concreteness . availability . misperception of frequency . selective perception . salience
COMPUTATIONAL (enhancing the capability to perform computations)	. limited computational ability . sequential processing . inconsistency . conservatism
PROCEDURAL (supporting implementation and documentation of a formal analytical process)	. misconception of frequency, uncertainty and chance . feedback errors (logical fallacies, hindsight, irrelevant learning structures) . schema
REPRESENTATIONAL (to extract as much information as possible from a display)	. functional fixedness . data presentation . scale effects . cognitive style . framing

Computational support provides the user with algebraic and data manipulation capabilities. Examples are handheld calculators, slide rules, and any device or formula that relieves the user from performing long and convoluted calculations unaided.

Process or procedural support provides the user with an explanation of the relevant decision rules, "normative" heuristics, models and decision processes. Examples are policy manuals, standard operating procedures, documentation of programs, help menus and displays, process tracers to document the analysis that has been performed, and any material that supports the user in clarifying analytical approaches to address the problem at hand.

Finally, display or representational support provides the user with different ways of displaying and formatting the data in an attempt to extract as much information as possible from that data. Examples are graphics packages and physical models, and any means of representing data in different and vivid ways.

Cognitive Support can be further specified. Indeed, the types of Cognitive Support can be matched to the cognitive limitations and biases they most effectively address. Table 3 shows some specific match-ups (for a detailed explanation of the specific biases see [1]). The idea is to identify what biases may hinder the problem solving process, and then design a system to minimize cognitive distortion by integrating the four types of cognitive support aimed at neutralizing those biases.

Organizational Support

Organizational support consists of developing an organization-wide information system based on a set of contingencies to switch information from one organizational unit to another, to involve certain organizational units in the decision making process and exclude others, and to establish redundant sources of information and channels of communication, etc. [14, 19, 20]. Organizational Support can be characterized by the extent to which it addresses what Huber [14] has called the "logistical determinants of performance and behavior" that define communication patterns among organizational units. That is, Organizational Support concentrates on improving the collection and transmission of the information required for decision making.

Organizational Support is determined by analyzing the communication system as a whole and identifying where in the system there may be opportunities for delaying, modifying, rerouting, or noxiously altering information. Organizational support then consists of (1) the design of the best routing and medium of communication for a particular message, and (2) the development of incentives and a reporting structure to minimize distortion. In other words, while Cognitive Support attempts to alleviate biases and limitations of the user, Organizational Support aims at reducing distortion by assuring the reliability and validity of the communication system.

Organizational Support consists, in a way, of building a communication road system. One must decide where to put highways (i.e., high speed links protected with passwords and where fees may be charged), where to put "ramps" onto the highway (i.e., who in the organization should be served by and have access to such high speed link), where to put two-lane roads (e.g.,batch processing) or even dirt roads (e.g., internal non-electronic mail).

Deciding on the architecture of t..e communication system requires an understanding of the historical communication patterns in the organization, and the specific relationship in terms of power, status, competition, workload, and distance between organizational units. Organizational Support implies that for each item of information a cost/benefit analysis should be performed to determine what sort of safeguards need to be implemented to insure a distortion-free transmission. The safeguards consist of switches determining special routings for specific messages (e.g., by-passing a unit that may have a reason to suppress the message; shortening the number of links), creating redundant channels, increasing the capacity and speed of specific channels, changing the medium of transmission, creating congruence of goals across organizational units, and implementing incentives to reduce distortion.

Decision Phases

If we are to improve decision making, one must understand the decision that needs to be made. A useful paradigm is to look at decision-making as a process consisting of a series of distinct and identifiable phases [22, 23]. In breaking the decision process into specific phases, one can analyze much more effectively what biases may affect each phase (e.g., in problem recognition we may want to make sure that availability and faulty recall are addressed), what are the crucial aspects of each (e.g., convergence or divergence), and whether there is a particularly useful technique or methodology (e.g., if the phase is the generation of alternatives, what is a good technique).

The degree of structure that each of the phases should have is important to determine because it leads to specific types of support processes, e.g., programs vs. analogies, optimization vs. heuristics. Whether the phase is well-structured or not also affects the degree of involvement of the user in developing the support system, the skills of the analyst and the type of information needed [24].

The framework to analyze DSS that is proposed here and shown in Table 4, is thus based on the following set of questions: (1) What are the decision phases that need to be/are being supported?, (2) What cognitive biases can affect those phases and how are they being dealt with?, and (3) What information is needed/being provided and what safeguards have been implemented to protect the reliability of that information?.

The framework can be thought of as a classification scheme to compare DSS in terms of the type of support they provide for a given problem. Attempting to compare DSS in terms of effectiveness of support is a difficult task. Alter [21] built a taxonomy based on the "degree of action implication of the system output", i.e., the degree to which the output can directly determine the decision to the problem it is being used on. As a taxonomy, the framework shown in Table 4 can be used to compare DSS on the **extent** of support they provide and underscores the importance of supporting the process as well as the output.

The framework can also be used as a "grading sheet" to determine the type of support needed. This determination can serve to guide the design of a new DSS or to predict and understand the performance of an existing one.

264

Two actual systems have been developed using this framework: One was developed to support career planning [25] and the other guided the development of an intelligent worksheet modeling system [26]. In both instances, the framework identified the critical biases to counteract and the type of support required to do so. Both systems, before being fully implemented, were tested in controlled experiments and shown to be remarkably successful in improving decision quality.

TABLE 4.
A Framework for Providing Decision Support

		DECISION PHASES		
TYPE OF SUPPORT NEEDED		PHASE 1	PHASE i	PHASE n
	INDEXING			
COGNITIVE	COMPUTATIONAL			
	PROCEDURAL			
	REPRESENTATIONAL			
TYPE OF DELIVERY NEEDED				
	REROUTING			
ORGANIZATIONAL	REDUNDANCY			
	CAPACITY/SPEED OF CHANNELS			
	GOAL CONGRUENCE			
	INCENTIVES			

THE DESIGN METHODOLOGY

The framework can be used as the basis for a design methodology. The methodology consists of nine broad steps organized into two categories: "Understand the context",(steps 1-6), and "design the system" (steps 7-9).

Understand The Context

Step 1:   Identify and define the problem to be solved.

Step 2:   Determine the different decision phases required to solve the problem. These phases can be general phases, e.g., problem definition, implementation etc., or specific phases concerning the task at hand, e.g., determination of demand, assessment of supply, etc.. Breaking the problem into more specific phases facilitates the identification of support and information needs.

Step 3: Determine the structure of each phase. Whether the phase is structured or not will determine, for example, whether normative models or heuristics are more appropriate in addressing the decision phase.

Step 4: Determine the potential cognitive biases affecting each decision phase. Analyze each information processing step and list what biases are likely to interfere.

Step 5: Determine the specific information items needed. This is a typical Information Requirement Determination step.

Step 6: Map the collection and transmission paths for each required information item. List the potential sources of distortion from collection through each transmission and storage step. Use Table 5 to classify distortions that have been identified. Determining whether the distortion is intentional or not helps choose the type of support required.

TABLE 5

Classification of types of distortion for each decision phase

TYPE OF DISTORTION	INTENTIONAL	NON-INTENTIONAL
COGNITIVE		
ORGANIZATIONAL		

Design The System

Step 7: Provide Cognitive Support to support the unbiased processing of information. Match the four generic forms of Cognitive Support to specific biases affecting each and all of the identified phases.

Step 8: Provide Organizational Support to assure distortion free collection and transmission of relevant information. Develop a strategy from the components of Organizational Support to assure the undistorted communication of all required information items.

Step 9: Integrate into one system.

An example: A system to support estimations

Prediction and estimation are common managerial tasks and can serve as an example to demonstrate the use of the proposed design strategy. The example is shown in Table 5 (to keep the example simple, organizational support is not addressed in detail).

Let us assume that a manager has to predict next year's demand for a given product (refer to column 1 in Table 6). Using a lens model approach (27) we can identify in five phases the task of predicting (column 2). Next, we identify the potential biases affecting each phase (column 3).

266

For example, when deciding what aspect of demand to estimate, the manager is likely to choose what he has always chosen in the past without critically evaluating the potential differences between next year and past years. When the manager tries to identify what indicators might be good predictors of demand, he may suffer from illusory correlation or use the most salient cues in his mind. When assessing the actual levels of those indicators, he may be "anchored" on a misleading value. In combining the indicators he might not be able to use them all or even use them consistently. All these potential pitfalls would affect his prediction.

TABLE 6
Designing a System to Support Prediction

(1) Identify Problem	(2) Phase	(3) Bias	(4) Type of Support			
			I	C	P	R
	Identify subject of prediction	. Schema	X		X	
		. Habit			X	X
	Identify predictive cues	. Availability	X		X	
		. Illusory correlation	X	X		
		. Frequency	X			
		. Selective perception			X	
		. Forced coherence	X		X	
	Assessment of levels	. Anchoring and adjustment	X		X	X
		. Representativeness	X			X
		. Unstated assumption			X	
	Combination of cues	. Inconsistency		X		
		. Conservatism		X		
	Feedback	. Misconception of chance	X	X	X	X
		. Regression to the mean	X	X		
		. Success/ failure attributions	X			
		. Gambler's fallacy	X		X	
		. Hindsight	X			

Key: I — Indexing; C — Computational; P — Procedural; R — Representational

From an organizational distortion point of view, we need to explore whether there are some rewards based on specific levels of demand that may induce the manager to over- or under-estimate demand, and whether other people which are part of the information network have a particular interest in this manager under- or over-estimating demand. Assuming that some individuals may want to inflate some of the figures needed by the manager, then redundant sources of information will have to be found and/or incentives for correct estimates introduced.

The next step then is to consider the type of support that should be provided to neutralize those biases (column 4). For example, a methodology or a series of questions (procedural support) directed at exploring the environment, the competitors, government regulations, etc., could be used to overcome potentially dysfunctional aspects of habit, selective perception and schema. A regression model (computational support) could help the manager in combining the different variables that the manager has chosen to use as predictors. Keeping a simple "score box" (indexing support) could avoid hindsight by documenting the processes that led to the predictions.

CONCLUSION

The field of DSS is permeated by a view that lack of information is the major obstacle to good decision making. This article proposes that the real limitations reside in poor processing and distorted transmission of the relevant information for decision making.

Indeed, distortion in communication and cognitive biases introduce errors in decision making. We suggest that the purpose of DSS is to minimize those errors. That is, their design should be guided by two considerations: a) to provide cognitive support to individuals, and b) to guarantee the integrity of organizational communications.

Furthermore, providing decision support requires a detailed understanding of the decision making process from both the descriptive and normative perspectives. In order to improve a decision process one must first define, analyze, and understand it. One cannot improve what one does not understand and breaking the decision process into specific phases is critical to that understanding.

Finally, we propose a methodology to design DSS consisting of five steps: (1) understanding what the decision is all about, and what process describes that decision best; (2) identifying the sources of cognitive bias that may affect the decision maker in each specific phase of that process; (3) designing the appropriate support to overcome those biases; (4) identifying the information needed to make the decision and how it will get to the decision maker; and (5) designing the appropriate organizational channels to obtain and deliver that information in the most reliable way. Parts of this methodology have been successfully applied to the development of a career planning system and to an intelligent worksheet modeling program. Further testing is on the way.

REFERENCES

1. Hogarth, R.M., _Judgment and Choice_, New York: John Wiley, Second Edition, 1987.

2. Mintzberg, H., "Managerial Work: Analysis from Observation," _Management Science_, Vol. 18, No. 1, 1971.

3. Mason, R. and Mitroff, I. "A Program for Research in Management Information Systems," _Management Science_, Vol. 19, No. 5, 1973.

4. Keen, P.G.W. and Scott-Morton, M.S., _Decision Support Systems: An Organizational Perspective_, Reading, MA: Addison-Wesley, 1978.

5. Sprague, R. H., and Carlson, E. D., _Building Effective Decision Support Systems_, Englewood Cliffs, NJ: Prentice-Hall, 1982.

6. Tversky, A. and D. Kahneman, "Judgment Under Uncertainty: Heuristics and Biases," _Science_, Vol. 185, September 1974.

7. Einhorn, H.J. and R.M. Hogarth, "Confidence in Judgment: Persistence of the Illusion of Validity," _Psychological Review_, Vol. 85, No. 3, 1978.

8. Fischhoff, B., "Perceived Informativeness of Facts," _Journal of Experimental Psychology_, Vol. 3, No. 3, 1977.

9. Driver, M. and Streufert, "Integrative Complexity. An Approach to Individuals and Group as Information Processing Systems," _Administrative Science Quarterly_, Vol. 16, No. 5, 1969.

10. Brightman, H.J. and T.F. Urban, "The Influence of the Dogmatic Personality Upon Information Processing: A Comparison with a Bayesian Information Processor," _Organizational Behavior and Human Performance_, Vol. 11, No. 2, 1974.

11. Taylor, R.N., and M.D. Dunnette, "Relative Contribution of Decision Makers Attributes to Decision Processes," _Organizational Behavior and Human Performance_, Vol. 12, No. 2, 1974.

12. Guilford, J.P., _The Nature of Human Intelligence_, New York: McGraw-Hill, 1967.

13. Janis, I. L. and Mann, L. _Decision Making: A Psychological Analysis of Conflict, Choice and Commitment_, New York: Free Press, 1977.

14. Huber, G.P., "Organizational Information Systems: Determinants of Their Performance and Behavior," _Management Science_, Vol. 28, No. 2, 1982.

15. Wiksell, W., _Do They Understand You?_, New York: McMillan, 1960.

16. MacCrimmon, K.R., "Descriptive Theory of Team Theory: Observation, Communication and Decision Heuristics in Information Systems," _Management Science_, Vol. 20, No. 10, 1974.

17. Athanassiades, J. C., "The Distortion of Upward Communication in Hierarchical Organizations," _Academy of Management_, Vol. 16, No.2, 1971.

18. Hackathorn, R. D., "DSS and Personal Computing," M.I.T. Center for Information Systems Research, No.47, 1978.

19. Galbraith, J.R., _Designing Complex Organizations_, Reading, MA: Addison-Wesley, 1973.

20. O'Reilly, C.A. and L. Pondy, "Organizational Communication," in S. Kerr (Ed.), *Organizational Behavior*, Columbus, OH: Grid, 1980.

21. Alter, S., *Decision Support Systems: Current Practice and Continuing Challenges*, Reading, MA: Addison-Wesley, 1980.

22. Kepner, C. and Tregoe, B., *The Rational Manager*, New York: Prentice Hall, 1982.

23. Humphreys, P., "Intelligence in Decision Support," in B. Brettner, et al (eds.) *New Directions in Research on Decision Making*, Amsterdam: North Holland, 1986.

24. Gorry, G.A. and Scott-Morton, M., "A Framework for MIS," Sloan Management Review, Vol. 11, No. 4, 1971.

25. Cats-Baril, W. and Huber, G., "Decision Support Systems for Ill-Structured Problems: An Empirical Study," *Decision Sciences*, Vol. 18, No.3, 1987.

26. Fordyce, K. J., "Looking at Worksheet Modeling Through Expert Systems Eyes," in B.G. Silverman (ed.) *Expert Systems for Business*, Reading, MA: Addison-Wesley, 1987.

27. Brunswick, E., *Perception and the Representative Design of Experiments*, Berkeley, CA: University of California Press, 1956.

SUPPORTING EXPERT JUDGEMENT OF HUMAN PERFORMANCE AND RELIABILITY

D.E. Embrey, Human Reliability Associates Ltd.,
1 School House, Higher Lane,
Dalton, Wigan,
Lancashire, WN8 7RP, UK

ABSTRACT

This paper begins by discussing an expanded role for human reliability assessment. The requirements for supporting analytical decision making in various aspects of human reliability assessment are discussed, and examples of techniques to provide this support are given.

INTRODUCTION

Over the last twenty years, most of the resources available for the development and application of the human reliability discipline have concentrated on one very narrow area: the quantification of human error probabilities for use in fault trees and safety analyses. In the opinion of many practitioners, the real goal of human reliability assessment should be improving the safety of potentially hazardous systems by the cost effective reduction of human error. From this perspective the emphasis on error probability quantification is a very inefficient use of resources. An expanded role for human reliability assessment is needed, which places its major emphasis on the following areas:

a)   The minimisation of human error by considering it in a systematic manner during the design of systems.

b)   The development of human reliability auditing and assessment techniques which emphasise the identification of critical errors and the development of strategies to minimise these errors.

c)   The analysis of operational data on human performance problems to allow the underlying causes of these problems to be evaluated and appropriate remedial measures to be implemented.

d) The development of a systematic framework, or the use of existing structures within the organisation (e.g. Quality Assurance Systems) to ensure the effective implementation of a human error minimisation programme.

e) The development of methodologies to address the effects of organisational structures and management policies on the safety and reliability of large technical systems.

Although the latter two areas can be seen as strategic objectives, significant progress has been made in the first three, and several techniques which have already been developed for other purposes can be modified for use in these applications. The remainder of this paper will concentrate on the first two areas, since the analysis of operational experience from the point of view of deriving underlying causes for human reliability problems has already been discussed in some detail in a publication by another attendee at this conference, Dr Deborah Lucas (see Lucas 1988). The areas of design and assessment both involve complex decision making and judgements on the part of the analyst. The next section of the paper will therefore address the needs for decision support in these areas. Subsequently, methods available for providing this support will be described and the further research needs discussed.

REQUIREMENTS FOR DECISION SUPPORT IN HUMAN RELIABILITY ASSESSMENT

Design

In the area of design, a prime requirement is a methodology that will assist the designer in making trade-offs between various design parameters in order to achieve a design which meets its functional specification. Another need is to demonstrate that a design will reliably fulfil its specification when exposed to a wide range of operational conditions. This need is only partially met by existing approaches such as user trials and simulations, since they only address a very limited subset of these conditions. A method is needed to generalise from the simulation to the real world. Another critical need is for a design tool which enables cost considerations to be related to the reliability and functionality of a design. Examples of other areas where decision support is required are the following:

- When generating alternative design options to meet the functional specifications and reliability requirements under all operating conditions.

- During cost-benefit analyses to evaluate the most cost effective option.

- Analysis of the cost implications of achieving a required reliability level.

Once a particular design option has been chosen, it is necessary to carry out a qualitative human reliability assessment, in order to identify any significant error modes and to modify the design to minimise the likelihood of their occurrence. The assessment methodology that is employed at this stage of the design process is identical to that used to audit an existing system. Finally, a quantitative human reliability assessment will be performed if this forms part of the specification. A systematic framework is necessary to provide decision support for both qualitative and quantitative analyses.

Human Reliability Assessment of Existing Systems

This form of human reliability assessment can be regarded primarily as an audit process to ensure that all significant error modes are identified and that suitable error reduction strategies are prescribed and implemented. The methodologies to achieve these objectives will be identical to those utilised during the design process to identify and eliminate potential errors. An audit assessment will utilise any analyses generated at the design stage together with any operational data indicating performance deficiencies that have arisen from causes not considered during the original human reliability analyses. The decision support requirements will be similar to those required during design. Cost considerations will still need to be taken into account when prescribing error reduction strategies.

TECHNIQUES AVAILABLE FOR DECISION SUPPORT OF HUMAN RELIABILITY ASSESSMENT

This section discusses examples of techniques which can be used to provide decision support for both design and human reliability auditing. They are by no means the only techniques that could be employed, but they illustrate the usefulness of a systematic framework to assist the analyst.

Design Applications

The technique illustrated in this context is an application of Multiattribute Utility Theory (MAUT) to assist the designer in choosing between design alternatives in terms of cost and reliability. This theory underlies the Success Likelihood Index Method (SLIM) technique for human reliability quantification, and the software which was developed for SLIM is also used in this application.

The basic premise of SLIM is that experts are able to develop a model which relates the effects of Performance Influencing Factors (PIFs, formerly called Performance Shaping Factors PSFs) such as quality of design, level of training, task complexity, time available etc. to the likelihood of task success. This likelihood is expressed in the form of an index called the Success Likelihood Index (SLI) which is in turn inversely related to the probability of failure.

When using the technique, judges first decide on a generic set of PIFs that determine the success likelihood for a set of tasks to be evaluated. Each task is then rated on each PIF. The generic importance weight of each PIF in terms of its effect on any task in the category being assessed is then evaluated using a computerised trade-off process (the weights are independent of the ratings). The SLI for each task is then calculated as the sum of the products of the ratings for each PIF and the weight assigned to the PIF.

As an example, the PIFs that determined success for a set of tasks might be design, time available and level of training. Two tasks might be assigned ratings on the PIFs as set out in table 1. (where the rating

TABLE 1: Calculation of SLI

PIF	task 1 rating	task 2 rating	PIF weights
design	7	3	0.2
time	2	2	0.3
training	5	9	0.5

scales go from 1 to 9). Thus, for task 1 a 7 indicates that the quality of design is good whereas 2 on the time scale indicates that only a short time is available to perform the task. The PIF weights indicate that training (weight 0.5) has a greater generic effect than design (weight 0.2). The SLI is calculated for task 1 and 2 as follows:

$$SLI_1 = 7 \times 0.2 + 2 \times 0.3 + 5 \times 0.5 = 4.5$$

$$SLI_2 = 3 \times 0.2 + 2 \times 0.3 + 9 \times 0.5 = 5.7$$

Thus, task 2 is more likely to succeed than task 1. Essentially SLIM allows the development of a model connecting the likelihood of a task succeeding (or failing) to the factors that influence that likelihood. In this simplified example this relationship is of the form:

$$SLI = 0.2 \ R1 + 0.3 \ R2 + 0.5 \ R3$$

where R1, R2 and R3 are respectively evaluations of the quality of design, time available to perform a task, and the level of training and experience possessed by the operator.

In a real application the design PIF would be decomposed further in terms of other more specific factors.

Even in this simplified example, it is possible to see how this methodology can be used for a variety of decision support purposes:

a) Deciding on the PIFs influencing reliability and the trade-offs between these PIFs. Using the sophisticated software package SAM (SLIM Assessment Module), the designer is provided with guidance in deciding on relevant PIFs and evaluating the relative effects that these exert on the tasks being assessed.

b) Evaluating alternative designs on the basis of costs. The latest version of SAM allows cost functions to be associated with each PIF rating. Thus, alternative designs, which may have similar SLIs and therefore similar expected reliability, could be differentiated in terms of costs.

c) Determining the costs associated with achieving a given reliability. Given cost functions and constraints on some of the PIFs, the costs associated with achieving a required reliability can be evaluated.

d) Generalising from simulations. The existence of a relationship between the success likelihood and the PIFs allows the results of trials or simulations to be generalised to a wide range of conditions

274

by changing the ratings on PIFs. For example, the expected reliability at different levels of training or experience could be assessed by changing R3.

Identification of Error Modes

A framework for the systematic identification of error modes has been developed which is a significant advance on earlier approaches such as SHERPA (Embrey, 1985) and the systematic search strategy of Pedersen (1984). This framework comprises the following stages:

a) Task analysis: An extension of the Hierarchical Task Analysis technique is used (Shepherd, 1984).

b) Error mode generation: Error modes are generated by considering the error inducing factors associated with each task. Planning and communication failures are considered in addition to action errors.

c) Screening procedure: A significance index for each error mode is calculated on the basis of its likelihood of occurrence, the likelihood of its recovery prior to a significant consequence, and the severity of the consequence. Error modes below a user defined cut-off are not evaluated further. For all errors above the cut-off an error minimisation strategy must be defined.

d) Error minimisation strategy definition: For all significant errors, the system assists the analyst in the development of an appropriate error reduction strategy by reference to the predominant error inducing factors.

This framework will ultimately be implemented in the form of a computer program for use during both the design and human reliability auditing stages.

CONCLUSIONS

As discussed at the beginning of this paper, human reliability as a discipline will only make a significant impact on the safety of technical systems if it extends its range beyond the narrow domain of human error quantification. This paper has discussed some of the ways in which this extension may be achieved, and has suggested how existing techniques can be extended to provide the basis for the methodologies required in these expanded areas of application.

REFERENCES

1. Embrey, D.E. (1986), SHERPA: A Systematic Human Error Reduction and Prediction Approach. Proceedings of the International Topical Meeting on Advances in Human Factors in Nuclear Power Systems, Knoxville, Tennessee, U.S.A.

2. Lucas, D.A. (1988), Human Performance Data Collection in Industrial Systems, in: Proceedings of a conference on human reliability in nuclear power. IBC plc, 3rd Floor, Bath House, 56 Holborn Viaduct, London, EC1A 2EX.

3. Pedersen, O.M. (1984), <u>Human Risk Contributions in Process Industry</u>, Report no. Risø-M-2513. Risø National Laboratory, DK-4000, Roskilde, Denmark.

4. Shepherd, A. (1984), <u>Hierarchical Task Analysis and Training Decisions</u>, Programmed Learning and Education Technology 22.2 162.